# STRENGTHENING OF CERAMICS

# MANUFACTURING ENGINEERING AND MATERIALS PROCESSING

*A Series of Reference Books and Textbooks*

SERIES EDITORS

### Geoffrey Boothroyd

*Department of Mechanical Engineering*
*University of Massachusetts*
*Amherst, Massachusetts*

### George E. Dieter

*Dean, College of Engineering*
*University of Maryland*
*College Park, Maryland*

1. Computers in Manufacturing, *U. Rembold, M. Seth, and J.S. Weinstein*
2. Cold Rolling of Steel, *William L. Roberts*
3. Strengthening of Ceramics: Treatments, Tests, and Design Applications, *Henry P. Kirchner*

*OTHER VOLUMES IN PREPARATION.*

# STRENGTHENING OF CERAMICS

## Treatments, Tests, and Design Applications

**Henry P. Kirchner**

*Ceramic Finishing Company*
*State College, Pennsylvania*

## CRC Press

Taylor & Francis Group
Boca Raton  London  New York

CRC Press is an imprint of the
Taylor & Francis Group, an **informa** business

First published 1979 by Marcel Dekker, Inc.

Published 2019 by CRC Press
Taylor & Francis Group
6000 Broken Sound Parkway NW, Suite 300
Boca Raton, FL 33487-2742

© 1979 by Taylor & Francis Group, LLC
CRC Press is an imprint of Taylor & Francis Group, an Informa business

First issued in paperback 2019

No claim to original U.S. Government works

ISBN-13: 978-0-367-45205-6 (pbk)
ISBN-13: 978-0-8247-6851-5 (hbk)

Visit the Taylor & Francis Web site at
http://www.taylorandfrancis.com

and the CRC Press Web site at
http://www.crcpress.com

Library of Congress Cataloging in Publication Data

Kirchner, Henry Paul.
    Strengthening of ceramics.

    (Manufacturing engineering and materials processing ; 3)
    Bibliography: p.
    Includes index.
    1. Ceramic materials.  I. Title.  II. Series.
TA455.C43K57          666              79-17514
ISBN 0-8247-6851-5

# PREFACE

Considerable resources have been devoted to diverse attempts to strengthen ceramic materials. In the majority of cases the results achieved have been marginal, leading to considerable frustration. Based on earlier success with strengthening of glass and glass-ceramics, using approaches involving treatments to induce compressive surface stresses, substantial progress was made in strengthening conventional polycrystalline ceramics and oxide single crystals. This monograph describes these advances. The methods described were developed, in large part, at Ceramic Finishing Company, but results from other laboratories that have come to my attention are included. The scope of this monograph is limited to treatments for polycrystalline ceramics and oxide single crystals. Despite the limited scope it was necessary to omit much important data. This information is available in the references.

Designers will find that, in addition to the descriptions of individual treatments and the resulting improvements in strength, information on potential applications, limitations of the treatments, design considerations, and costs are presented.

The principal advantages of compressive surface layer treatments are improved strength, decreased penetration of surface damage and strength degradation, and improved impact, thermal shock and delayed fracture performance. The highest strengths have, thus far, been achieved by thermal (quenching) treatments. These treatments are limited, to some extent, by thermal shock damage in the larger and more complex shapes. Coatings, chemical treatments and treatments to induce phase transformations are alternative processes. Although the achievable strengths using these treatments are somewhat lower, a much wider range of shapes and sizes can be treated.

Potential applications should be sought in the following areas: bearings, cutting tools, gas turbines, radomes, IR domes, armor, extrusion dies, pump parts, lighting envelopes, laser windows, leading edges, other wear resistance parts, and so forth. Progress in the

application of "tempered" glasses and glass-ceramics can provide some
guidance as to what to expect. It should be noted that the payoff
for improvements in many of these applications may be very large
because the ceramic part is the critical link in the operation of the
complete device. Operations of complete machines may succeed or fail
depending on the success or failure of these individual ceramic parts.
In these circumstances, substantial improvements in strength may be
essential to reliable performance.

Subcritical crack growth is the principal factor affecting the
reliable use of many ceramics in load bearing applications. The
original flaws gradually increase in size to the point that stress
intensification may cause catastrophic failure. To reduce subcritical crack growth, one can reduce the load or redesign the part so that
the stress is less. However, in many cases neither means is satisfactory. Use of compressive surface stresses provides another alternative. These surface stresses, because they subtract from normal
tensile stresses in the surface where most of the severe flaws are,
reduce the stress intensity factor at flaws in stressed members,
thus decreasing the subcritical crack growth rate.

Proof testing is an accepted means for assuring the strength of
ceramic parts by removing parts weaker than a particular strength
level from the distribution. The most critical proof test parameter
is the difference between the proof test stress and the use stress.
The greater the difference can be, the greater will be the effectiveness in removing weak parts from the lot, especially when the subcritical crack growth during loading and unloading from proof testing
is considered. Because the compressive surface stresses raise the
nominal stresses at which surface flaws act to cause failure, treated
specimens can be proof tested at stresses that are much higher than
otherwise possible. For a given use stress this means a larger difference between the proof stress and the use stress and therefore a
smaller probability of failure of a part in service.

More widespread use of ceramics can aid greatly in improvement
of efficiencies of energy conversion equipment and in weight reduction in aerospace applications. Improvements in energy efficiency
can occur by means such as increased inlet temperatures leading to
increased Carnot efficiencies and by reduction in the requirements
for cooling air in gas turbine engines. Weight reduction can be
accomplished by substitution of ceramics for refractory metals, reduction of the thickness of sections through the use of the greater
stiffness of some ceramics, and by other means. Weight reductions
in aerospace applications can be especially important because less
fuel is needed leading to still further weight reductions. The increased strength and reliability of ceramics with compressive surface
stresses will aid in adapting ceramics to these applications.

It is a pleasure to acknowledge the contributions of my assoc-
iates at Ceramic Finishing Company, especially Robert M. Gruver and
others, many of whom have been coauthors of the various earlier descriptions of this work, and the many research workers at other
organizations who have contributed to our knowledge of this field.
I am also grateful to the sponsors of the research including especially the Naval Air Systems Command and the Office of Naval Research,

and the technical contract monitors, Charles F. Bersch and Arthur M.
Diness.  Helpful suggestions were made by Richard C. Bradt and
John B. Wachtman, Jr. who reviewed the original manuscript.  It is
also a pleasure to acknowledge the typing done by Shirley Wittlinger
and Elaine Smiles.

Henry P. Kirchner<br>
April, 1979

CONTENTS

Preface                                                          iii

Chapter 1.  Introduction                                           1

    1.1.  The Strength of Ceramics                                 1
          1.1.1.  Stress concentrations at flaws                   2
          1.1.2.  Griffith's theory                                3
          1.1.3.  Fracture mechanics                               4
          1.1.4.  Flaw characteristics                             6
          1.1.5.  Relationship of strength and grain
                  size                                             8
          1.1.6.  Environmental effects and slow crack
                  growth                                           9
          1.1.7.  Strength distributions                          11
          1.1.8.  Strengthening mechanisms                        12

    1.2.  Strengthening by Compressive Surface Stresses           13
          1.2.1.  Criteria for selection of bodies for
                  strengthening                                   14
          1.2.2.  Stress profiles                                 17

    1.3.  Characterization and Property Measurements              20
          1.3.1.  Ring test                                       20
          1.3.2.  Rod test                                        21
          1.3.3.  Fracture mirror measurements                    22
          1.3.4.  Indentation test                                23
          1.3.5.  Flexural strength                               24
          1.3.6.  Tensile strength                                26
          1.3.7.  Impact resistance                               28
          1.3.8.  Thermal shock resistance                        29
          1.3.9.  Penetration of surface damage                   29

Chapter 2.   Thermal Treatments                                    31

        2.1.  Strengthening Sintered Alumina by Quenching          31
              2.1.1.  Flexural strength of 96% alumina
                      quenched into various media                  32
              2.1.2.  Glazing and quenching                        36
              2.1.3.  Tensile strength                             45
              2.1.4.  Effect of quenching temperature on
                      surface forces and flexural strength         51
              2.1.5.  Flexural strength distributions              54
              2.1.6.  Elevated temperature flexural
                      strength                                     56
              2.1.7.  Impact resistance                            60
              2.1.8.  Thermal shock resistance                     63
              2.1.9.  Delayed fracture                             65
              2.1.10. Strength degradation caused by
                      surface damage                               73
              2.1.11. Stress profiles                              74
              2.1.12. Other sintered alumina bodies                79

        2.2.  Strengthening Hot Pressed Alumina by
              Quenching                                            81
              2.2.1.  Flexural strength                            82
              2.2.2.  Elevated temperature flexural strength       86
              2.2.3.  Penetration of surface damage                91
              2.2.4.  Stress profiles                              95
              2.2.5.  Flaws at fracture origins                   102

        2.3.  Strengthening Sapphire by Polishing,
              Reheating and Quenching                             110
              2.3.1.  Flexural strength                           113
              2.3.2.  Thermal shock resistance                    117
              2.3.3.  Analysis of strength and thermal
                      shock test data                             117

        2.4.  Strengthening Other Oxide and Silicate
              Ceramics by Quenching                               120
              2.4.1.  Titania ($TiO_2$)                            121
              2.4.2.  Spinel ($MgAl_2O_4$)                         121
              2.4.3.  Steatite ($MgSiO_3$)                         124
              2.4.4.  Forsterite ($Mg_2SiO_4$)                     126
              2.4.5.  Zircon porcelain                             128
              2.4.6.  Electrical porcelain                         128
              2.4.7.  Mullite                                      128

        2.5.  Strengthening Non-Oxide Ceramics by
              Heating and Quenching                               132
              2.5.1.  Strengthening silicon carbide by
                      heating                                      132
              2.5.2.  Strengthening silicon carbide by
                      quenching                                    134

2.5.3.  Strengthening silicon nitride by
        heating                                          137
2.5.4.  Strengthening silicon nitride by
        quenching                                        139

2.6.    Status of Research on Thermal Treatments         145
        2.6.1.  Summary of successes and failures        148
        2.6.2.  Effect of various shapes and sizes
                on strengthening results                 148
        2.6.3.  Summary of benefits of quenching         153
        2.6.4.  Recommendations for research             154


Chapter 3.  Coatings and Chemical Treatments             155

3.1.    Glazes                                           155
        3.1.1.  Low expansion glazes                     155
        3.1.2.  Strengthening glazed alumina by
                ion exchange                             158

3.2.    Low Expansion Solid Solution Surface Layers      163
        3.2.1.  Polycrystalline alumina                  167
        3.2.2.  Sapphire                                 173
        3.2.3.  Polycrystalline titania                  176
        3.2.4.  Polycrystalline spinel (MgAl$_2$O$_4$)   177
        3.2.5.  Polycrystalline magnesia                 182
        3.2.6.  Periclase (MgO)                          186
        3.2.7.  Polycrystalline nickel oxide             187
        3.2.8.  Silicon nitride                          188

3.3.    Low Expansion Compound Surface Layers Formed
        by Reaction with the Surface                     188
        3.3.1.  Polycrystalline alumina                  188
        3.3.2.  Sapphire                                 199
        3.3.3.  Polycrystalline magnesia                 203

3.4.    Low Expansion Coatings                           203

3.5.    Phase Transformations                            205

3.6.    Status of Research on Coatings, Chemical
        Treatments and Phase Transformations             208
        3.6.1.  Summary of successes and failures        209
        3.6.2.  Summary of the advantages of
                compressive surface stresses induced
                by coatings, chemical treatments and
                phase transformations                    211
        3.6.3.  Recommendations for research             212

Chapter 4.  Potential Applications                                                    215

    4.1.  Applications                                                                215

    4.2.  Limitations of Treatments                                                   215

    4.3.  Design Considerations                                                       216
          4.3.1.  Subcritical flaw growth                                             216
          4.3.2.  Scatter in the strengths                                            217
          4.3.3.  Surface damage                                                      218
          4.3.4.  Efficiencies of energy conversion
                  equipment                                                          219
          4.3.5.  Weight reduction                                                    220

    4.4.  Costs                                                                       220

References                                                                           223

Index                                                                                237

# CHAPTER 1

## INTRODUCTION

### 1.1 THE STRENGTH OF CERAMICS

If the strength of ceramics were determined solely by the
stress necessary to separate planes of atoms, one would expect
ceramics to be very strong. However, high strength is observed
infrequently and only in special materials (Kelly, 1966). In
almost all cases, ceramics are weaker than otherwise expected
because flaws concentrate the applied stresses.

In a few cases, flaw free or almost flaw free specimens can
be prepared. Sapphire single crystals grown by the Czochralski
technique are relatively free of internal flaws. If these crystals
are subsequently flame polished, the surfaces can be made flaw free
and the flexural strengths may exceed 10,000 MPa (1,450,000 psi)
(Kelly, 1966). This strength is a substantial fraction of the
theoretical strength of the material. Silica fibers can be prepared to yield comparable strengths. The high strengths that have
been achieved in these special cases provide goals to be achieved
in further development of other ceramics. The recent book of Lawn
and Wilshaw (1975) is a good source of general information about
the strength and fracture of brittle solids.

1

### 1.1.1  Stress Concentrations at Flaws

The magnitudes of stresses concentrated at flaws can be
estimated based on the initial work of Inglis (1913).  A hole of
elliptical cross section passing through a thin plate subjected
to uniform tensile stress applied perpendicular to the major axis
of the ellipse is shown in Figure 1.1.  The maximum stress is the
tensile stress in the y direction at point A which can be calculated
using

$$\sigma_y = \sigma \left[ 1 + 2 \left(\tfrac{a}{\rho}\right)^{1/2} \right] \qquad\qquad (1.1)$$

in which $\sigma$ is the uniform applied stress, 2a and 2b are the lengths
of the major and minor axes respectively and $\rho$ is the radius of
curvature at A (Kelly, 1966).  For small radii of curvature, the
stress concentration factor is large so that the first term in
parenthesis can be neglected leading to

$$\sigma_y \cong 2\sigma \left(\tfrac{a}{\rho}\right)^{1/2} \qquad\qquad (1.2)$$

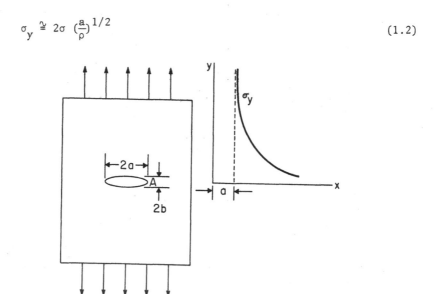

Figure 1.1.  Stresses at an elliptical hole through a flat plate.

In attempting to apply this equation to real cracks it has sometimes
been assumed that $\rho$ is one half of the interatomic distance (so-
called atomically sharp cracks). The validity of this assumption
is doubtful because it leads to stress concentration factors of
about 1000, whereas values of 20-50 are more realistic based on
available fracture stress vs. flaw size data and estimates of the
theoretical strength.

The stress distributions at notches were analysed by Neuber
(1958). These calculated stress distributions have been confirmed
by photoelastic methods, in some cases.

### 1.1.2 Griffith's Theory

Griffith (1920) assumed that the minimum energy required to
fracture a body is the energy required to form the new surfaces.
The availability of this energy is a necessary but not a sufficient
condition for fracture to occur. For a flat, homogeneous, isotropic
plate of uniform thickness, containing a straight crack of length
2c passing normally through it, and subject to a tensile stress
applied at its outer edge, the fracture stress ($\sigma_f$) is (for plane
stress conditions)

$$\sigma_f = (\frac{2E\gamma}{\pi c})^{1/2} \tag{1.3}$$

in which E is Young's modulus and $\gamma$ is the thermodynamic surface
free energy (Cottrell, 1964).

To use the above equation to calculate the fracture stress,
it is necessary to measure the length of the naturally occurring
flaw at the fracture origin, or to form artificial flaws of known
length. Both have been done in a few cases. In practice it has
been found that the calculated fracture stresses are much too low.
Usually, the error is attributed either to the presence of a barrier
to crack initiation or to additional energy required to propagate
the crack over that needed to provide the thermodynamic surface
free energy of the new surfaces.

The work necessary to form the new surfaces during fracture
can be measured by several methods (Duga, 1969; Evans, 1974). The
resulting fracture energies ($\gamma_f$) are frequently one or two orders
of magnitude greater than the specific surface energy.  In metals
and other ductile materials most of this excess energy can be
accounted for by the energy absorbed by plastic flow processes.
Although there is evidence that similar processes occur in ceramics,
the energy absorbed is usually very small and cannot account for the
difference between specific surface energy and fracture energy.  For
example, Guard and Romo (1965) found that, for polycrystalline
alumina with an average grain size of 20 μm, the plastic work of
fracture is about 1500 ergs $cm^{-2}$ and the specific surface energy is
about 1000 ergs $cm^2$, whereas the fracture energy usually is found
to be in the range 15,000-66,000 ergs $cm^{-2}$.  Also, Wiederhorn,
Hockey and Roberts (1973) reported that plastic deformation by
dislocation motion or twin formation plays no role in the fracture
process in sapphire at temperatures below 400°C.  The absence of
the mechanism in the single crystal material leads one to question
that plasticity makes an important contribution to the fracture
energy of the polycrystalline material.  Other mechanisms that
contribute to the fracture energy are surface roughness and localized cracking.  Rough estimates of the contributions of these
mechanisms have been made in some cases but the major portion of
the energy difference remains unaccounted for.

The role of fracture energy in increasing resistance to fracture is widely recognized.  Research on methods of increasing
fracture toughness is underway in several laboratories and substantial increases in fracture toughness have recently been achieved.

### 1.1.3  Fracture Mechanics

Fracture mechanics is the study of forces, stresses and strains
at stationary and moving cracks.  An important factor in the development of fracture mechanics has been the realization that the need

for detailed knowledge of the processes occurring at the crack tip
can be avoided in many cases.  This has been done by introducing
the so-called stress intensity factor K and the effective surface
energy ($\gamma_e$) which are related by

$$K = (2E\gamma_e)^{1/2} = (GE)^{1/2} \qquad\qquad (1.4)$$

in which the so-called crack extension force G = $2\gamma_e$.  For cracks
opening under the influence of tensile stress (Mode I crack opening)
K is symbolized by $K_I$ which is given by (Evans and Tappin, 1972;
Brown and Srawley, 1966)

$$K_I = \frac{Y}{Z}\, \sigma a^{1/2} \qquad\qquad (1.5)$$

in which Y is a geometrical factor which adjusts for the relative
sizes of the crack and the specimen, Z is the flaw shape parameter
and (a) is a crack dimension, usually the crack depth.

The advantages of this approach can be illustrated by comparisons for blunt and sharp cracks.  A blunt crack of depth (a) will
fracture at a relatively high stress and based on (1.5) a relatively
high $K_I$.  Based on Equation 1.4 this high $K_I$ appears to be the
result of a high value of $\gamma_e$ even though bluntness has no physical
relationship to surface energy.  Similarly, a sharp crack causes
fracture at low stresses leading to a low value of $\gamma_e$.

Subcritical crack growth has similar influences.  For example,
if a crack or flaw of known shape grows along the surface and
becomes more severe, this fact should be compensated for by a
decrease in Z.  In practice, this change in shape is seldom discovered, the fracture occurs at a lower stress than otherwise
expected leading to a lower apparent $\gamma_e$.

The results of many fracture mechanics analyses have been
compiled by Tada, Paris and Irwin (1973).  This compilation is a
good starting point for application of fracture mechanics techniques to ceramics.

### 1.1.4  Flaw Characteristics

Because the stress concentrations at flaws are the principal
factor determining the strength of ceramics, it is important to
have information on the characteristics of flaws and the relationship of these characteristics to the strength.  In early investigations of flaws at fracture origins it was uncertain which one of
several flaws in the fracture surfaces was the flaw at which the
fracture originated.  Improvements in strength, finer grain size
materials and improved methods of fractography permit greater cer-
tainty in current determinations.  Kirchner, Buessem, Gruver, Platts
and Walker (1970) and Kirchner, Gruver and Walker (1973) studied
fracture origins in ceramics with and without compressive surface
stresses induced by quenching.  Evans and Tappin (1972) developed
methods for calculating fracture stresses at flaws, including pores
and machining defects, taking into account the possibilities of
flaw linking prior to catastrophic failure.  Kirchner and Gruver
(1973) used fracture mirror measurements to estimate fracture
stresses at internal fracture origins in hot pressed alumina.
Rhodes, Berneburg, Cannon and Steele (1973) characterized a small
number of flaws and correlated fracture stress and flaw size for
all types of flaws grouped together.  Rice and McDonough (1972 a,b)
and Rice (1974) studied flaws at fracture origins in a wide variety
of materials including spinel, zirconia, three types of alumina,
lead zirconate titanate, β-alumina and silicon nitride.  More
recently Gruver, Sotter and Kirchner (1976) and Kirchner, Gruver
and Sotter (1978) fractured several materials at various temperatures and loading rates.  The flaws at the fracture origins were
identified, characterized and classified into various types.  These
types included large crystals (surface and internal), pores (surface
and internal), stepped flaws and penetration flaws.  Stepped flaws
are found at or near the surface and have a steplike appearance.
Apparently, in this case the fracture initiates by linking up of
flaws in two or more different planes parallel to the fracture

surface. Penetration flaws probably arise because of surface damage in a plane that is intersected by the fracture surface. As a result, these penetration flaws appear in the fracture surface as linear defects extending from the surface toward the axis of the specimen. When a sufficient number of specimens were tested at a particular temperature and loading rate and failed at a particular type of flaw, fracture stress vs. flaw size curves were plotted and compared with theoretical curves.

Evans and Tappin (1972), Rhodes, Berneberg, Cannon and Steele (1973) and Gruver, Sotter and Kirchner (1976) have all emphasized that fracture stress in ceramics is controlled by the characteristics of the flaws rather than by the average microstructure. There are many examples of this. The example already given by sapphire crystals which after normal handling have strengths in the range 350-700 MPa, but which after flame polishing have strengths exceeding 10,000 MPA is striking. In polycrystalline ceramics several different types of flaws acting at nearly the same fracture stress are usually present. If such a material tends to fail at surface damage caused by machining, the fracture stress can be raised by removing the damage by lapping and polishing. However, in most ceramics the increases are rather limited because the origins are shifted to other types of flaws.

The strongest ceramics are usually single phase bodies or bodies in which the other phases are present as crystals or glassy regions that are very small relative to the crystals of the principal phase. The reason for this is that localized stresses are induced as a result of unequal thermal contractions of the phases during cooling from the sintering temperature and as a result of differences in the elastic constants of the phases. In brittle materials no mechanism is available to relieve these stresses so localized cracks may form. If these cracks are present the bodies are weak.

## 1.1.5  Relationship of Strength and Grain Size

Experimentally, it is well known that the strength of polycrystalline ceramics decreases with increasing average grain size
but the reasons for this grain size dependence are not well understood.  It may occur either because critical flaw sizes increase
with increasing grain size for many types of flaws or that the
single crystal fracture energy controls fracture when the flaw
size is much smaller than the grain size.  Some cases in which
this should be expected are the following:

1.  Fractures originating at large grains.  It is reasonable to
    expect that the size of the largest grains will increase with
    increases in the average grain size.  As shown by Kirchner,
    Gruver and Sotter (1976) there is a strong negative correla-
    tion of fracture stress and grain size of large grains at
    fracture origins.
2.  Fractures originating at localized cracks caused by thermal
    expansion anisotropy.  During cooling after sintering, local-
    ized stresses are induced by the unequal thermal contractions
    of the individual anisotropic crystals.  In many cases, these
    stresses are high enough to cause formation of localized cracks.
    The lengths of these cracks increase with increasing grain size
    (Kirchner and Gruver, 1970) leading to a related decrease in
    strength.
3.  Fractures originating at localized cracks caused by elastic
    anisotropy.  All crystalline bodies consist of elastically
    anisotropic cyrstals.  When these bodies are stressed, the
    stresses are transmitted unequally in various crystallographic
    directions, depending on the elastic anisotropy, leading to
    localized stresses.  If the lengths of cracks occurring as a
    result of these stresses increase with increasing grain size,
    as is the case for thermal expansion anisotropy, then the
    strength should also decrease.

In the case of pores and defects introduced by machining, the
relationships between flaw size and grain size, if any, are less
obvious.  Individual pores observed at fracture origins are usually
much larger than the grain size and there is no necessary correlation between pore size and grain size.  Fractures sometimes originate
at porous regions much larger than the grain size.  Tressler,
Langensiepen and Bradt (1974) reported a strong increase in strength

with decreasing grit size of diamond abrasives used to machine fine
grained alumina ceramics.  They concluded that machining defects
controlled the strength of the stronger fine grained ceramics but
that intrinsic flaws controlled the strength of coarse grained
material.  Kirchner, Gruver and Sotter (1976) showed that at least
two types of machining defects, stepped and penetration flaws, can
be much larger than the grain size and act as fracture origins in
dense, fine grained ceramics.

The situation is further complicated by the fact that determinations of the local stress at failure in cases where the failure
originates at isolated large grains or groups of large grains,
located internally, show that the stresses required to cause failure
are much greater than the strength usually measured for specimens
of approximately the same grain size (Kirchner and Gruver, 1973).
This observation indicates that the strength vs. grain size relations usually observed are surface related properties.

In summary, because of the many types of flaws, surface and
internal, there are many types of interactions between flaw size
and grain size.  The observed strength vs. grain size data depend
on the many types of interactions mentioned above so that no simple
relationship between strength and grain size should be expected.

### 1.1.6  Environmental Effects and Slow Crack Growth

Measurements of crack velocity vs. stress intensity factor at
various humidities have shown that, at low values of $K_I$, the crack
velocities of many ceramics increase with increasing humidity
(Wiederhorn, 1974).  As shown in Figure 1.2, these curves have
three regions; Region I in which crack velocity is reaction rate
limited, increases with $K_I$, and depends on the humidity, Region II
in which crack velocity is transport limited and is independent of
$K_I$ and Region III in which crack velocity increases with $K_I$ but is
independent of humidity.  The resulting slow crack growth leads to
substantial loading rate dependence of strength of many ceramics.
For example, decrease in fracture stress with decrease in loading

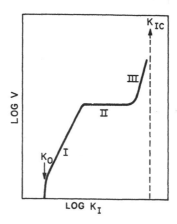

Figure 1.2.  Variation of stress intensity factor $K_I$ with crack
velocity V for a ceramic material in a corrosive
environment.

rate or during delayed fracture tests has been observed for alumina
by numerous investigators (Pearson, 1956; Williams, 1956; Charles
and Shaw, 1962; Dawihl and Klinger, 1966; Sedlacek, 1968; Sarkar
and Glenn, 1970).  The decrease in strength of alumina at various
humidities and loading rates is proportionately much less than that
of most silicate glasses.

Subcritical crack growth can also occur in absence of environmental effects as shown by Wiederhorn, Johnson, Diness and Heuer
(1974) for several glasses tested in a vacuum.  There is some
evidence that subcritical crack growth can occur in polycrystalline
ceramics in the absence of environmental effects.  For example, pores
at fracture origins in 96% alumina fractured at -196°C are surrounded
by a flat cleavage region that appears to be subcritical crack growth
(Kirchner, Gruver and Sotter, 1976).  Unless there is a reactive
gas in the pores and -196°C is not low enough to prevent chemical
reaction, both of which seem unlikely, such subcritical crack growth
occurs in the absence of a reactive environment.

It is clear from the fact that subcritical crack growth is such
a widespread phenomenon that many previous efforts to estimate

critical crack sizes from the sizes of pores, machining damage,
large crystals and other flaws have underestimated the critical
crack sizes. Efforts to remedy this problem depend on development of fractographic criteria for the boundary between subcritical
and critical crack growth. It is reasonable to expect that these
criteria can be developed.

### 1.1.7 Strength Distributions

When interest was focused on the strength of polycrystalline
ceramics a decade ago and widespread efforts were made to measure
the strengths, ceramic materials acquired a reputation for unreliability. There were a number of reasons for this including the
following:

1. Problems in adapting measurement techniques to ceramic materials.
2. Sensitivity of some of the stronger, fine grained materials to
   surface damage.
3. Variations within lots of materials caused by practical fabrication problems such as nonuniform distribution of grain growth
   inhibitor.

It was not uncommon to find groups of specimens in which the measured strengths of the weakest specimens were less than half those
of the strongest specimens.

One response to this problem has been to use very large numbers of specimens in an effort to obtain reliable estimates of
the strength distributions (Barnett, Costello, Herman and Hofer,
1965) (Pears, Starrett, Bickelhaupt and Braswell, 1970). This
approach is necessary in some cases, especially when the distribution of the strengths of the weakest specimens is needed with a
high degree of reliability. However, this approach is to a degree
self defeating because of difficulties in maintaining standardized
conditions during fabrication of a large group of test specimens,
assuring that the same standardized conditions can be maintained

during manufacture of subsequent groups, and assuring that the
data have a meaningful relationship to the performance of ceramic
parts with a variety of shapes and sizes and subjected to a variety
of stress states and environmental conditions.

Other responses to this problem are to reduce the variability
by decreasing the sensitivity to surface damage and to improve the
uniformity of the material.  Some methods that are effective in
implementing these responses are described in later sections.

### 1.1.8  Strengthening Mechanisms

Strengthening mechanisms that can be used to obtain higher
strengths in ceramics have been reviewed (Burke, Reed and Weiss,
1966).  A wide range of strengthening mechanisms or processes can
be identified including the following:

```
annealing
compressive surface stresses
dispersion strengthening
fiber reinforcement
minimizing grain size
minimizing porosity
reduction of crystal anisotropy by solid solution additions
reduction of localized stresses by enhancement of preferred
    orientation
solid solution hardening
strain hardening
unidirectional solidification
```

The effectiveness of flame polishing of sapphire and etching
of glass for reduction of critical surface flaws has been stressed.
Nevertheless, reduction of critical surface flaws is impractical
in many cases because the material is subjected to surface damage
in use.  Many of the other methods have been used during the last
decade to obtain improved strength.  In most cases the observed
improvements have been unimpressive when the processes were applied
to the strongest available materials.  Exceptions are the results
obtained by using compressive surface stresses.  For example, the

room temperature flexural strength of dense, fine grained alumina was increased by about 100%. Improvements in thermal shock resistance, impact resistance, delayed fracture performance, and resistance to penetration of surface damage were achieved. Therefore, processes to form compressive surface stresses in ceramics were extensively investigated. The resulting processes and properties are described in this monograph.

## 1.2  STRENGTHENING BY COMPRESSIVE SURFACE STRESSES

Compressive surface stresses have been used to strengthen ceramics since antiquity. Most pottery and chinaware are glazed with glazes that have lower expansion coefficients than the bodies to which they are applied. As the materials are cooled to room temperature after glazing, the body tends to contract more than the glaze, placing the glaze in compression and the body in tension. The compressive glaze raises the nominal stress at which surface flaws act to cause failure, thus improving the strength.

In the decades since 1930, stronger ceramic bodies were developed, mainly by sintering of relatively pure oxide powders (Ryskewitch, 1960). The mechanism of fracture of these bodies was not well understood. In some cases it was assumed that fracture tended to originate at internal flaws rather than at surface flaws (Pears and Starrett, 1966). Based upon the uncertainty about the relative importance of surface flaws and internal flaws it was not always obvious that compressive surface layers would be effective in strengthening these newer oxide ceramics that were intrinsically so much stronger.

Another factor inhibiting the use of this approach to strengthening was the absence of obvious methods for inducing compressive surface stresses. Warshaw (1957) used the differential shrinkage of two different alumina porcelain bodies to form compressive surface layers but the strengths of the original ceramic bodies and the degree of improvement achieved were unimpressive.

Brubaker and Russell (1967), in research performed in 1962 and earlier, formed two-layer laminated ceramics using three vitrified whiteware bodies having different thermal expansion coefficients, leading to tensile and compressive stresses in the layers. The flexural strengths and impact resistances of the laminates were measured. Substantial improvements in both properties were observed when the compressive layers were subjected to tensile forces due to the externally applied loads. Even so, the resulting bodies were not very strong in comparison with readily available fine-grained oxide ceramics.

Progress in the last decade has, to some degree, clarified the roles of surface flaws and volume flaws in the fracture of ceramics and has made available a number of techniques for inducing compressive surface stresses in strong ceramic bodies. Among these techniques are the following:

1. Quenching
2. Glazing
3. Glazing and quenching
4. Ion exchange glazing
5. Forming low expansion (high expansion) solid solution surface layers
6. Forming low expansion (high expansion) compound surface layers
7. Forming surface layers by reactions or phase transformations characterized by an increase in volume

It is evident that if compressive stresses are desired at elevated temperatures, this can be achieved by applying higher expansion layers to lower expansion bodies. The rate of relaxation of the stresses at elevated temperatures should be evaluated. These processes will be described in detail in the later sections.

## 1.2.1  Criteria for Selection of Bodies for Strengthening

The principal criterion for selection of bodies for strengthening by compressive surface stresses is that fracture due to externally applied loads must normally originate at surface flaws rather

than at internal flaws.  If fracture normally originates at internal
flaws the strength will not be improved and, if substantial internal
tensile stress is induced to offset the compressive surface stresses,
the body may even be weakened.

To determine that the fractures normally originate at the
surface is difficult in many cases.  It is especially difficult in
relatively weak bodies because the fracture surfaces are rather
flat and featureless.  Jacobsen and Fehrenbacher (1966) measured
the flexural strengths of 37 rectangular bars of hot pressed magnesia,
13 μm average grain size, and observed individual strengths ranging
from 23,700 to 34,800 psi.  Except for a few cases in which fracture
originated at edge chips, careful examination of the specimens
yielded little information about the characteristics of critical
flaws.  Heuer (1969) compared the flat, featureless surface of
alumina fractured after being weakened by a notch and the surface
of a specimen fractured at high stresses.  The area in which the
fracture origin is located is evident in the stronger specimen.  In
the surfaces of specimens fractured at low stresses, several flaws
may be observed but usually there is no definitive evidence to
indicate which one of the flaws served as the fracture origin.

Strong, fine grained ceramic bodies have fracture surfaces
with well defined fracture features.  These features include a
mirror region, mirror boundary and hackle.  These fracture features
can be observed by breaking a piece of ordinary glass rod and
examining the fracture surface at magnifications of 10X to 30X.  An
illustration of such a fracture surface is given in Figure 1.3, in
which the fracture mirror is visible as the smooth region surrounding the fracture origin at the top of the picture.  The hackle are
the radiating ridges and valleys.

The factors that determine whether or not these fracture features are observed in a particular case are not completely understood.  The fracture stress, elastic constants and fracture surface
energy have important influences.  Also, the specimen size is important.  In some cases the mirror boundary is not observed because the

Figure 1.3.   Fracture mirror in a flint glass rod quenched from
              725°C into forced air and fractured in flexure.


required stress intensity is not attained within the dimensions of
the specimen but the mirror boundary would be observed if the specimen were larger.  Large grained bodies have two disadvantages.
Usually they are weak, leading to poorly defined fracture features.
Even if they are strong the fracture surfaces have a granular
appearance that may obscure needed fracture features.

     Three methods are suitable for locating fracture origins in
ceramics (Gruver, Sotter and Kirchner, 1976).


1.   Searching at the intersection of the hackle, extended through
     the fracture mirror.
2.   Searching at the intersection of the fracture mirror radii.
3.   Searching in the region of reflecting spots (if present).


     In some investigations, the observation that the distribution
of the individual strengths can be fitted to a Weibull distribution

or that the average strength decreases with increasing specimen
volume have been cited as evidence for a volume dependence of
strength.  These are unreliable methods.  When it is possible to
do so, the relative frequencies of fracture origins at the surface
and in the interior should be determined and the method chosen for
analysis of the data should be based on these results.

In addition to the general criterion that the body must fracture at surface flaws, there are many other criteria that apply to
selection of treatments or processes to be used to form compressive
surface stresses in specific bodies.  Using the list of processes
in Section 1.2 as a guide, one can suggest some of these factors.
For example, treatments involving quenching must start with bodies
having sufficient thermal shock resistance, those involving low
expansion surface layers must not have expansion coefficients so
low that suitable low expansion compositions for the surface layers
are unavailable; and so forth.

### 1.2.2  Stress Profiles

Some ways in which the stress profiles in the treated specimens depend on the type of treatment will be illustrated briefly
in this section.  The detailed methods used to determine the stress
profiles will be described along with the specific treatments in
later sections.  It is important to emphasize that the description
of the stress profiles presented here is simplified in that only
the axial stresses are accounted for and the radial and circumferential components are neglected.

The differences in the stress profiles depend mainly on the
relative thickness of the compressive surface layer.  Thin compressive
surface layers usually give rise to low or moderate axial tensile
stresses in the interior.  Therefore, it is not likely that these

internal stresses will interfere with the strength.  Examples of
such thin surface layers are illustrated in Figure 1.4 (A) and
(B).  Effects of applied flexural and tensile loads on the stress
profiles are indicated.

The thicker surface layers, usually obtained by quenching, give
rise to greater residual tensile stresses in the interior as illustrated in Figure 1.4 (E).  These residual tensile stresses may
combine with the applied loads to cause failure in some specific
circumstances; particularly if internal flaws are present.

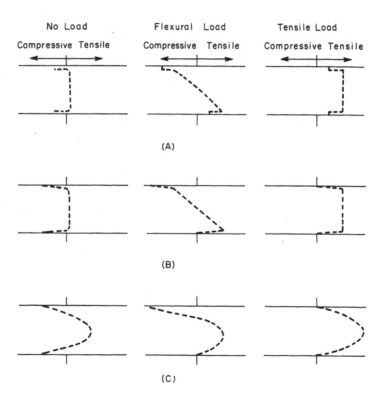

Figure 1.4.   Schematic stress profiles.   (A) Uniform stress in thin
              coating.   (B) Graded stress in a diffused surface layer.
              (C) Stress profile in quenched specimen.

It is important to recognize that the principal mechanism
involved in strengthening of ceramics by quenching is the increase
in the nominal stress at which surface flaws act to cause failure
as a result of the presence of the residual compressive surface
stresses.  This mechanism is analogous to that involved in strengthening of glasses by quenching ("tempering" and "thermal tempering"
are the terms customarily used to designate quenching of glasses).
It is a completely different mechanism from that usually involved
in strengthening of metals by quenching, in which case one makes
use of the presence of non-equilibrium phase transitions to enhance
the strength.

It is necessary to understand the changes that occur during
quenching in order to optimize quenching conditions successfully.
To achieve this understanding, it is best to think in terms of the
strains involved and when necessary to convert these strains into
stresses.  These changes in strain and stress have been discussed
in detail by Weymann (1962) and Buessem and Gruver (1972).  During
the initial stages of rapid cooling, the surface cools much more
rapidly than the interior.  Therefore, the surface tends to contract leading to tensile stresses in the surface and compressive
stresses in the interior.  In response to these stresses, relaxation occurs in the surface and in the interior but because the
interior is at higher temperatures the relaxation in the interior
can be expected to be greater than that at the surface.  The
relaxation acts to relieve the stresses and they are at least
partially relieved before the relaxation rates become small at low
temperatures and relaxation ceases contribution to the strain.  At
this point the interior is still at higher temperature than the
surface and during subsequent cooling the interior contracts more
that the surface overcoming whatever residual thermal tensile
strain remained from the period during which relaxation was
occurring.  As a result there is a stress reversal with the
surface going into compression and the interior into tension.

## 1.3  CHARACTERIZATION AND PROPERTY MEASUREMENTS

Every technological advance is dependent upon and limited by the techniques that can be applied to the particular subject matter.  The literature abounds with examples of studies that failed to achieve their objectives because suitable techniques were unavailable.  Therefore, it has been necessary to develop or modify methods of evaluation and characterization.  These developments have been carried out along with development of the treatment processes and are described in this section to reduce confusion in the remainder of this monograph.

### 1.3.1  Ring Test

A fundamental problem is to detect and measure the compressive surface forces.  There are many ways to do this but in practice it is difficult to find a practical method.  The first method used in in the present work was the so-called ring test, which is a modification of the test described by Schurecht and Pole (1930).  This test involved treating the outside surface of a hollow cylinder, cutting a section from the wall and measuring the movement of the surfaces of the cut as the restraint was removed.  If compressive surface layers were present, the surfaces of the cut tended to move closer together.  The magnitude of the movement, compared with similar specimens, was a measure of relative residual surface force.

The ring test provided useful information but was not very sensitive and, in some cases, the treatment affected both the inside and outside surfaces.  Another difficulty was that, frequently, ring shaped pieces were not available.

### 1.3.2 Rod Test

In an effort to remedy some of the deficiencies of the ring test, the rod test was developed (Gruver and Kirchner, 1968). In this test the diameter of a treated cylindrical rod (or width of a rectangular bar) was measured and, then, the rod was slotted axially as shown in Figure 1.5. After slotting, the diameter was measured again at the tip of the slot. The original diameter was subtracted from the diameter of the rod after slotting. For similar specimens this difference in diameters is a measure of the relative residual surface force. With some additional information, the surface stress can be calculated in simple cases (Gruver and Buessem, 1971). If the rod tips moved closer together compressive surface forces were present. If they moved farther apart, tensile surface forces were present.

In some untreated materials, the rod tips tend to move apart after slotting. Usually, this movement occurs because of damage to the surfaces of the slot during sawing. Another possibility is

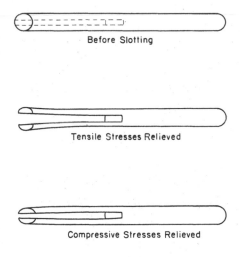

Before Slotting

Tensile Stresses Relieved

Compressive Stresses Relieved

Figure 1.5 Rod test.

that residual tensile stresses were present in the surfaces of the
untreated rod.  Frequently, it is useful to determine a correction
factor by slotting several untreated rods and averaging the results.
Subtracting this correction factor from the results obtained for
treated rods gives a more reliable measure of the surface forces.
Because the correction factor is usually negative, subtraction of
the correction factor usually yields a higher estimate of the
relative surface force.

The choice of the rod diameter, slot width and slot length
should be made with care (Gruver and Buessem, 1969; 1971).  One
problem is that for thin surface layers and low stress levels the
moment exerted by the compressive surface layer is small and the
deflection may be small and insensitive to changes in stress level.
This problem can be overcome by increasing the slot width and slot
length.  Another problem is that the deflection may be so large that
slot tips touch for all of the treatments of interest.  In this case
increasing the slot width simply makes the prongs more flexible.
Therefore, it may be necessary to decrease the slot length or
increase the rod diameter to remedy this problem.

### 1.3.3  Fracture Mirror Measurements

The fracture stress ($\sigma_f$) is generally accepted to be related
to the fracture mirror radius ($r_m$) by the following relation
(Terao, 1953; Orr, 1972)

$$\sigma_f r_m^{1/2} = A \qquad\qquad (1.6)$$

in which A is a constant for a particular material tested under a
particular set of test conditions.  The appropriate fracture stress
for use in this expression is the local stress.  Therefore, if
large scale residual stresses are present, the fracture mirror
radius can be used, together with the above equation, to determine
the local stress ($\sigma_L$) and the nominal tensile stress ($\sigma_N$) can be

calculated from the applied loads using the methods of linear
elasticity. Then, the residual stress ($\sigma_R$) can be determined
using the following equation (Orr, 1972).

$$\sigma_L = \sigma_N + \sigma_R \tag{1.7}$$

If the residual stresses at the surface are compressive, as desired,
the local stress is less than the nominal stress. Thus, the sign
convention to be used in the above equation is that tensile
stresses are positive and compressive stresses are negative. The
units of A in Equation 1.6 are stress intensity units. Congleton
and Petch (1967) investigated crack branching in alumina, magnesia
and glass and expressed their observations in terms of the stress
intensity factor at crack branching ($K_B$) as

$$K_B = Y \, \sigma_f C_B^{1/2} \tag{1.8}$$

in which Y is the geometrical factor suitable for the crack at
branching. Since that time it has become customary to analyse crack
branching and fracture mirror boundary formation in terms of stress
intensity or closely related criteria (Kirchner, 1978).

### 1.3.4  Indentation Test

A very recent development is the use of the indentation test
to estimate residual surface stresses (Marshall and Lawn, 1978;
Marshall, Lawn, Kirchner and Gruver, 1978). Combining fracture
mechanics equations for indentation damage penetration and remaining strength yielded as theoretical expression for the residual
surface stresses. The characteristic radius (C) of a well
developed, half penny shaped crack produced by indentation, for
example, by a Vickers pyramid with an indentation load (P) is

$$\frac{P}{C^{3/2}} = (\frac{K_{IC}}{X})(1 + \frac{2}{\pi^{1/2}} \, \sigma_R C^{1/2}) \tag{1.9}$$

where $\sigma_R$ is the residual compressive stress, $K_{IC}$ is the critical
stress intensity factor and $\chi$ is a dimensionless indenter constant.
For surfaces in which such a crack is the critical flaw, the subsequent fracture stress is

$$\sigma_F = \frac{\pi^{1/2} K_{IC}}{2C^{1/2}} + \sigma_R \qquad\qquad (1.10)$$

Elimination of the crack length from the above equations yields

$$P = \frac{\pi^{3/2}}{8} \left(\frac{K_{IC}}{\chi}\right)^4 \frac{\sigma_F}{(\sigma_F - \sigma_R)^4} \qquad\qquad (1.11)$$

$K_{IC}$ and $\chi$ can be determined by calibration experiments on residual
stress free specimens by solving Equations (1.9) and (1.10) simul-
taneously with $\sigma_R = 0^*$. Subsequent experiments on specimens with
compressive surface stresses yield estimates of $\sigma_R$.

It is useful to distinguish between strength increases caused
by flaw healing that may occur during treatment to induce compressive surface stresses and those due to the residual stresses. The
strengthening effect of the treatment can be expressed as

$$\sigma_T - \sigma_0 = (\frac{\pi^{1/2}}{2} K_{IC})(C_T^{-1/2} - C_0^{-1/2}) + \sigma_R \qquad\qquad (1.12)$$

in which the subscripts T and 0 refer to the treated and as received
surfaces. At high loads the indentations are the critical flaws and
the strengthening effect tends toward $\sigma_R$ but if healing occurs the
strengths at low loads may be greater than otherwise expected.

### 1.3.5  Flexural Strength

Appropriate methods of measurement of flexural strength have
been discussed at length in reports of government sponsored research
(Rudnick, Marshall, Duckworth and Emerick, 1968; Pears, Starrett,
Bickelhaupt and Braswell, 1970; Bortz and Wade, 1967). Usually,

---
[*] In specimens free of large scale residual stresses, localized
residual stresses at the indentation may reduce the fracture stress.

these tests are done by either three point or four point loading.
The choice between these two methods depends on the objectives of
the investigation.  Ideally, three point loading leads to maximum
stresses only under the central load point whereas, four point
loading leads to maximum stresses over the distance between the
central load points.  Since the probability that a flaw of a particular degree of severity will be encountered in the area subjected
to maximum stress depends on that area, three point loading yields
higher average strengths and greater scatter in the results than
four point loading.  Therefore, if one is interested in how much
strength improvement might be achieved in bodies with reduced flaw
densities, it may be reasonable to use three point loading but if
one wants conservative engineering data or small scatter to improve
the reliability of comparisons, it is best to use four point loading.

   The principal errors in flexural tests are:

1. Friction forces under the load points.  These forces give rise
   to bending moments that oppose those due to the applied load,
   thus increasing the load required to cause failure.  These
   forces can be reduced by using rollers as the load points.
2. Distortion of the stress distribution due to "wedging."  "Wedging"
   is used to describe the local crushing forces at the load points.
   The effect of these forces can be held to reasonable levels by
   using a large enough span to depth ratio.  If the span to depth
   ratio is in the range 8-10, reasonable results are usually
   obtained.  If the ratio is as low as 5 or 6, caution should be
   used before accepting the data.  If it is necessary to use span
   to depth ratios that are too small, it is desirable to express
   results as a fraction or percentage of control values in order
   to prevent unrealistically high strength data from accumulating
   in the literature.
3. Distortion of the stress distribution due to twisting.  These
   forces give rise to parasitic stresses that are not accounted
   for in the analysis so that lower loads are required to cause
   failure and low strengths are observed.  These errors can be
   avoided by careful machining of specimens, careful alignment
   of the test fixture and by using cylindrical rods instead of
   rectangular bars as test specimens.
4. Incorrect spacing of load points and unequal distribution of
   loads.  In four point loading, this condition leads to larger
   stresses at one of the interior load points than at the other.
   This error can be minimized by using a swivel in the fixture
   holding the inner load points and by accurate construction and
   alignment of the test fixture.

Some ceramics are known to be susceptible to stress corrosion
and others may be susceptible.  Therefore, it is important to consider the effect of the test environment on the test results.  In
some cases it may be necessary to determine the effect of humidity
and stressing rate on the strength.  In other cases it may be sufficient to make the measurement under controlled humidity conditions.

Over the last 14 years during which this work was done, there
have been a very rapid changes in the customary units used to express
results of mechanical property measurements.  These changes have
occurred at different rates in various disciplines with the literature in some areas completely converted to SI units while other
areas are still using English and other metric units.  Because of
this confusion, it was decided to retain the units in which the
original measurements were made, in this monograph.  Thus, in most
cases strengths are reported in pounds per square inch (psi) or
megapascals (MPa or $MNm^{-2}$).  In some important cases the results
are given in alternate units in parenthesis.  Conversion can be
done using one MPa or $MNm^{-2}$ = 145 psi.

### 1.3.6  Tensile Strength

The factors to be considered in making good tensile tests on
brittle materials have been described (Rudnick, Marshall, Duckworth and Emrick, 1968; Pears, Starrett, Bickelhaupt and Braswell,
1970; Bortz and Wade, 1967).  The principal problem is to assure
accurate alignment of the grips and axial application of the load
in order to avoid superimposing bending or torsion (parasitic)
stresses on the intended uniform tensile stress.  Several tests
including the Stanford ring test (Sedlacek and Halden, 1962), the
gas bearing tensile test (Pears and Starrett, 1966) and a test
using a thermal contraction loading mechanism (Kirchner and Rishel,
1971) were developed in attempts to deal with this problem.

The apparatus based on the thermal contraction loading
mechanism was used in investigations of the tensile strength of

ceramics with compressive surface stresses. The absence of long
screws and moving parts, the rigidity of the apparatus, the potting
of the specimens in the grips help to minimize parasitic stresses.
Some results of tensile and flexural strength measurements are
presented in Table 1.1 and Figure 1.6 (Platts and Kirchner, 1971).
The specimens were cylindrical rods of a 96% alumina ceramic that
were necked down by grinding to form the test section. Normally
these specimens are used only for tensile testing but in this case
some of the specimens were fractured in flexure. The average tensile
strength was lower than the average flexural strength. This result
was expected because, in tensile testing, the surface area subjected
to the maximum stresses is much greater than in the flexural test
so that the likelihood that relatively severe flaws will be in the
area of maximum stress is increased. Nevertheless, the tensile
strength is a large fraction of the flexural strength reflecting
the relative absence of parasitic stresses. The scatter of the
test results and the distribution curves are similar in both sets
of data. Because parasitic stresses are likely to be variable
in magnitude, they are likely to increase the scatter. Therefore,
it seems that in this case parasitic stresses did not make a major
contribution to the observed difference in the averages.

Table 1.1. Comparison of Tensile and Flexural Strengths of 96%
Alumina Specimens in As-Ground Condition

|                          | Flexure Data | Tension Data |
| ------------------------ | ------------ | ------------ |
| No. Specimens            | 19           | 11           |
| Average Strength-psi.    | 49,100       | 39,800       |
| Standard Deviation-psi.  | 3,500        | 3,980        |
| Coef. of Variation       | 7.1%         | 10.0%        |
| Ratio of Strengths       | 0.81         |              |

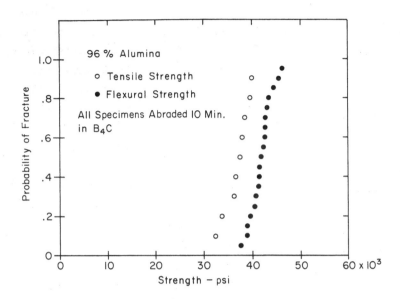

Figure 1.6.  Comparison of distributions of tensile and flexural
             strengths of lightly abraded 96% alumina.  Reprinted by
             permission of the American Society for Testing and
             Materials, Copyright, 1971.

### 1.3.7  Impact Resistance

The resistance to impact was measured by the drop weight and
Charpy methods.  In the drop weight method a steel ball was dropped
from increasing heights onto the center of a rod supported at both
ends.  The energy at which failure occurred was used as the measure
of impact resistance.

Charpy tests were performed using a BTL type impact testing
machine manufactured by Satec Systems, Inc.  This machine was modi-
fied for elevated temperature impact testing by building an induction
heated furnace and specimen supports between the pendulum supports.
The scatter of impact test results was determined for specimens
supported on a 1.5 in. span in the furnace.  The test specimens were
96% alumina rods, 0.3 in. in diameter, with as-fired surfaces.  The
coefficient of variation observed for a group of nine measurements

was 11.5%. Therefore, this method is capable of distinguishing differences in impact resistance for groups of specimens having substantially different impact resistances.

### 1.3.8 Thermal Shock Resistance

Thermal shock resistance has been evaluated by many methods. In the present case the resistance of the material to surface crack initiation due to thermally induced tensile stresses near room temperature was measured. The presence of the surface cracks was detected by measuring the remaining flexural strength after one thermal shock cycle. The surface cracks were induced by quenching the specimens from a furnace into water at room temperature. The temperature of the furnace was increased by increments until the remaining strength of the specimens decreased sharply. The thermal shock resistance was expressed in terms of the temperature change that the specimens withstood before this damage occurred.

### 1.3.9 Penetration of Surface Damage

One of the principal advantages of the use of compressive surface layers is to prevent or reduce degradation of the strength as a result of abrasion or other surface damage. Compressive surface layers reduce penetration of surface damage. The compressive surface layers also reduce strength degradation by raising the nominal stress at which these surface flaws act to cause failure.

Several techniques have been used to study the effect of compressive surface layers on penetration of surface damage. These techniques include the following:

1. Inducing controlled damage and measuring the effect of this damage on the strength of treated and control specimens.
2. Microscopic examination of the damage in the surface of the piece as, for example, examination of the track formed by a loaded diamond point drawn across the surface.

3.  Microscopic examination of the fracture surface of damaged
    specimens to observe the original damage and evidence of damage
    propagation.
4.  Microscopic examination of penetration of surface damage in the
    plane perpendicular to the damage.  For example, a loaded
    diamond point may be drawn across the surface.  Then, the
    piece may be notched on the opposite side and fractured back
    toward the diamond scratch.  In this case the fracture surface
    shows the diamond scratch intersecting the fracture surface and
    damage penetration may be observed under the scratch.  Scanning
    electron microscopy is an appropriate technique for observing
    this damage penetration.

# CHAPTER 2

## THERMAL TREATMENTS

### 2.1 STRENGTHENING SINTERED ALUMINA BY QUENCHING

It is well known that rapid cooling can improve the strength
of various ceramics (Hummel and Lowery, 1951; Smoke and Koenig,
1958; Dunsmore, Fenstermacher and Hummel, 1961; Philips and
DiVita, 1964).  In these references, the strengthening process
involved cooling the specimens in an air blast and it was usually
called thermal conditioning.  Increases in strength up to 111% were
reported but, because the starting materials were not very strong,
the resulting strengths often remained well below the strengths of
the strongest available ceramics.

In more recent research, stronger materials were used as the
starting materials.  They were cooled rapidly by more effective
liquid and gaseous quenching media.  Present evidence indicates
that specimens quenched from temperatures at which the body is
slightly plastic are more resistant to thermal shock than would
normally be expected.  Therefore, the specimens survive much more
rapid cooling than would have been thought possible.  Using these
improved materials and methods, substantial improvements in strength
were achieved.  These improvements are accompanied by improvements
in impact resistance, thermal shock resistance, delayed fracture
performance, and resistance to penetration of surface damage.

Because of their availability and reasonable cost, commercially
available 96% alumina ceramics were used to develop these processes.

Subsequently, these processes were applied to fine grained, hot
pressed alumina to obtain higher strengths.  Development of these
processes, applied to the 96% alumina bodies, is described in the
following sections.

### 2.1.1  Flexural Strength of 96% Alumina Quenched into Various Media

96% alumina[*] rods were quenched into various media including
silicone oils[+] of various viscosities, and a variety of other liquid
and gaseous media (Kirchner, Walker and Platts, 1971).  The resulting flexural strengths are given in Table 2.1.  These results
demonstrate that substantial improvements in flexural strength can
be achieved by quenching into either liquid or gaseous media, with
the highest values observed for several liquid media.  The best
improvements were about 120% and were achieved by quenching into
oils of low viscosity.  Water and ethylene glycol were so effective
in cooling the alumina rods that thermal shock damage and strength
degradation occurred.

Silicone oils are desirable quenching media for several
reasons, especially the substantial thermal stability of these
compounds.  There are several disadvantages (some of which are
shared with other quenching media) including cost and the fact that
decomposition of the oil results in deposition of a dark brown
surface layer on the pieces.  One way to reduce these disadvantages  is to use emulsions as quenching media.  Despite the fact
that water by itself usually causes thermal shock damage, emulsions containing as much as 99% water have been used successfully.

---

[*]ALSIMAG #614, American Lava Corp., Chattanooga, Tenn., composition
approximately 1.5% MgO, 2.5% $SiO_2$ and 96% $Al_2O_3$ average grain size
approximately 5μm.

[+]Dow Corning No. 200 dimethylpolysiloxane.

Table 2.1.  Flexural Strength of 96% Alumina Quenched into Various
            Media

| Quenching Medium | Quenching Temp. °C | No. Specimens | Average Flexural Strength[*] psi |
|---|---|---|---|
| Liquid Media | (0.125 in diam rods) | | |
| As-received controls | --- | 19 | 46,800 |
| Silicone oil (5.0 centistokes) | 1550 | 5 | 102,100 |
| Silicone oil (5.0 centistokes) | 1600 | 5 | 105,600 |
| Silicone oil (100 centistokes) | 1550 | 5 | 105,200 |
| Silicone oil (100 centistokes) | 1600 | 5 | 105,400 |
| Silicone oil (100 centistokes) | 1650 | 5 | 95,100 |
| Silicone oil (350 centistokes) | 1600 | 3 | 86,000 |
| Silicone oil (1000 centistokes) | 1600 | 3 | 91,000 |
| Silicone oil (12500 centistokes) | 1550 | 5 | 86,100 |
| Motor oil (SAE 30) | 1600 | 5 | 102,100 |
| Kerosene | 1600 | 5 | 87,100 |
| Corn oil | 1500 | 5 | 76,800 |
| Olive oil | 1500 | 5 | 92,400 |
| Castor oil | 1500 | 5 | 99,800 |
| Cod liver oil | 1500 | 5 | 85,400 |
| Linseed oil | 1500 | 5 | 94,300 |
| Triethanol amine | 1500 | 5 | 84,100 |
| Lard at 100°C | 1500 | 5 | 94,100[+] |
| Ethylene glycol | 1500 | 5 | 17,400[+] |
| Water | 1600 | 3 | 5,200[+] |
| Gaseous Media | (0.137 in diam rods) | | |
| As-received controls | --- | 19 | 47,900 |
| Refired controls, 1500°C | --- | 5 | 59,600 |
| Forced air | 1500 | 5 | 77,400 |
| Forced helium | 1500 | 5 | 80,100 |
| Forced $CO_2$ | 1500 | 5 | 76,900 |

[*]Four point loading on a two inch span.

[+]All specimens damaged by thermal stresses.

Dilution of the oil with water in various proportions leads to a
proportionate decrease in cost of the quenching medium.  In addition,
the presence of the water reduces decomposition of the silicone oil,
both on the surface of the alumina and in the bulk of the quenching
medium.  The heat of vaporization of the water may prevent the
temperature of the quenching medium from rising to the temperatures
necessary for decomposition of the silicone oil.

The flexural strengths of rods quenched into various emulsions
are listed in Table 2.2.  In the first series of emulsions, 2.5
grams of oleic acid and 1.5 grams of morpholine were added as
emulsifiers to each 100 grams of emulsion and mixed in a Waring
blender.  As indicated in the Table, mixtures containing as much
as 95% water were used successfully without thermal shock failure.
The highest average strength, 97,400 psi, was achieved by quenching
from 1550°C into an emulsion containing 5% silicone oil and 95%
water.  Quenching from higher temperatures resulted in lower
strengths.

Based on the above results, it seemed likely that the emulsifiers were affecting the cooling properties of the water.  Therefore, experiments were performed using quenching media in which
the concentration of emulsifier was decreased, then eliminated
entirely, and using mixtures of water and emulsifiers but no silicone oil.  These results are also listed in Table 2.2.  The concentration of the emulsifier was reduced by preparing the emulsion
containing 50% silicone + 50% water plus emulsifiers and then
diluting it with water alone.  For example, the resulting 25%
silicone oil + 75% water emulsion contained half as much emulsifier as the corresponding emulsion in the previous group.  Comparing
the various groups does not indicate a significant change in
flexural strength as a result of decreasing the concentration of
emulsifier.  When the emulsifier was eliminated completely, good
results were obtained with mixtures containing as much as 90%
water.  These mixtures were sufficiently stable for at least one
hour after mixing.  At higher percentages of water some thermal
shock failures were observed.

Table 2.2.  Flexural Strength of 96% Alumina Quenched into Various
Emulsions

| Quenching Medium | Quenching Temp. °C | No. Specimens | Average Flexural Strength* psi |
|---|---|---|---|
| As received | --- | 5 | 58,300 |
| Silicone oil-water emulsions with emulsifier | | | |
| 75% Silicone oil + 25% $H_2O$ | 1550 | 3 | 89,000 |
| 50% Silicone oil + 50% $H_2O$ | 1550 | 3 | 95,900 |
| 25% Silicone oil + 75% $H_2O$ | 1550 | 3 | 92,800 |
| 15% Silicone oil + 85% $H_2O$ | 1550 | 3 | 90,800 |
| 5% Silicone oil + 95% $H_2O$ | 1550 | 3 | 97,400 |
| Silicone oil-water emulsions Concentration of emulsifiers decreased | | | |
| 25% Silicone oil + 75% $H_2O$ | 1550 | 3 | 92,200 |
| 10% Silicone oil + 90% $H_2O$ | 1550 | 3 | 84,100 |
| 5% Silicone oil + 95% $H_2O$ | 1550 | 3 | 85,800 |
| 1% Silicone oil + 99% $H_2O$ | 1550 | 3 | 83,500 |
| Silicone oil-water emulsions with no emulsifier | | | |
| 75% Silicone oil + 25% water | 1550 | 4 | 84,500 |
| 10% Silicone oil + 90% water | 1550 | 4 | 92,900 |
| Water + emulsifiers | | | |
| 2.5 gms oleic acid + 1.5 gms morpholine per 100 gms mixture | 1500 | 3 | 90,600 |
| Same as above | 1600 | 3 | 88,800 |
| Same as above diluted 25% | 1600 | 3 | TSF+ |
| Silicone oil-water emulsions with a different emulsifier | | | |
| 75% Silicone oil + 25% water 3 gm stearic acid + 3 gm triethanolamine in 300 ml of medium | 1550 | 4 | 80,600 |

*Four point loading on a two-inch span.

+TSF-Thermal Shock Failures.

In mixtures containing only water and emulsifiers some good results were obtained with the original emulsifier concentration. However, on further dilution thermal shock failures were observed.

## 2.1.2  Glazing and Quenching

Glazing and quenching can be combined to obtain additional improvements in strength (Kirchner, Gruver, and Walker, 1968).  The mechanism of strengthening by glazing and quenching is probably not exactly the same as that involved in quenching alone.  One reason for this is the existence of a phenomenon that is observed in glassy materials but not in crystalline materials, namely that the specific volume depends on the cooling rate.  Therefore, during quenching the glaze is frozen into a low density structure so that the thermal contraction is less than expected.  As a result, the tensile stresses in the surface are lower during the early stages, less compressive creep occurs in the interior, and the resulting stresses depend to a greater degree than might be expected on the thermal contraction of the alumina and its concurrent compression of the glaze.

Many of the experiments with glazing and quenching were performed in the early stages of the research before it was discovered that ceramics could be quenched in appropriate liquids without thermal shock failure.  Therefore, most of the experiments with glazes involved quenching in forced air.

Compositions and thermal expansion coefficients of the glazes used to strengthen alumina are presented in Table 2.3.  Glazes were selected with thermal expansion coefficients above and below that of the alumina body.  The glaze used in most of the experiments is termed the "regular glaze."  This glaze has a coefficient of thermal expansion of $53 \times 10^{-7}$ °C$^{-1}$ for the temperature range of 25°-300°C.  In this same range a 96% alumina body has a coefficient of thermal expansion of $65 \times 10^{-7}$ °C$^{-1}$ giving a desirable degree of mismatch.

Table 2.3.  Compositions[*] and Thermal Expansion Coefficients of
            Various Glazes

|  | Regular Glaze | L-2 Glaze | S-1 Glaze |
|---|---|---|---|
| $SiO_2$ | 59.3 | 69.5 | 45.8 |
| PbO | 10.7 |  |  |
| $Al_2O_3$ | 12.0 | 5.1 | 31.0 |
| $B_2O_3$ | 4.4 |  |  |
| CaO | 8 | 12.5 |  |
| $Na_2O$ | 2.4 | 12.5 | 23.2 |
| $K_2O$ | 1.4 |  |  |
| Other | 2.3 |  |  |
| Thermal exp. coef., 25-300°C, $\times 10^{-7}$°$C^{-1}$ | 53 | 75 | 103 |

[*]Composition in weight percent.

The glaze compositions were compounded using typical glaze
raw materials.  The use of these materials is illustrated below
for the case of the regular glaze which was prepared by mixing the
following materials[+]:

|  | Weight Percent |
|---|---|
| G-24 frit | 56 |
| Nepheline syenite | 10 |
| Talc (tremolite) | 3 |
| Florida kaolin | 10 |
| Whiting | 7 |
| Flint | 14 |

The G-24 frit has the following composition:  $SiO_2$, 53.5%; $Al_2O_3$,
8.42%; $B_2O_3$, 7.44%; $ZrO_2$, 1.33%; $Na_2O$, 2.24%; CaO, 6.53%; MgO,
0.80%; PbO, 18.30%; $K_2O$, 1.41% and $Fe_2O_3$, 0.04%.  Three hundred

[+]Ceramic Color and Chemical Co., New Brighton, Pa.

grams of dry powder plus 150 ml of water were milled in a two
quart ball mill with alumina milling balls for approximately one-
half hour.

A slip was prepared by diluting the ball mill product with
water.  The slip was applied to the alumina specimens by dipping
or brushing.  The coated rods were fired and quenched using the
usual techniques.

## Characterization of the treated specimens

Electron microprobe analyses indicated that the glaze composition of the regular glaze changes considerably as a result of
firing at high temperatues.  Most of the lead was lost.  $B_2O_3$ was
not determined.  Adjusted to total 100%, the remaining major oxides
were present in the following percentages: $SiO_2$, 69; PbO, 2; $Al_2O_3$,
22 and CaO, 7%.

Relative residual surface forces in the glazed and quenched
rods were determined by ring tests and by slotted rod tests.  A
ring, glazed and quenched from 1500°C, closed 70 μm when slotted.
By comparison a control that was quenched but not glazed closed
only 14 μm.  Therefore, compressive forces were present in the
surfaces of all of the quenched specimens but when the specimens
were glazed and quenched the compressive forces were much larger.

The results of the rod tests are given in Table 2.4.  In the
as-received condition or when refired and slowly cooled the rods
opened when slotted indicating either the presence of tensile
surface forces or damage in the slot.  However, when the bare rods
were quenched the surface stresses changed from tensile to compressive.  Glazing with the regular glaze and slow cooling also resulted
in compressive surface stresses indicating that after evaporation
of the lead and reaction with the surface of the alumina body, the
thermal expansion coefficient of the surface layer remained less
than that of the alumina body.  As in the ring test, glazing and
quenching resulted in much greater compressive surface forces than
were observed in the other cases.

Table 2.4.  Rod Test Results (96% $Al_2O_3$ Rods, Quenched in Forced Air)

| Treatment | No. Specimens | Average Change in Rod Diam. in |
|---|---|---|
| Regular Glaze | | |
| Rod diam 0.15 in, slot 1.25 in long x 0.032 in wide | | |
| As received | 3 | +0.0024 |
| Cooled with kiln (1500°C, 1 hr) | 3 | +0.0027 |
| Refired and quenched (1500°C, 1 hr) | 3 | −0.0010 |
| Glazed and quenched (1500°C, 1 hr) | 3 | −0.0017 |
| Rod diam. 0.123 in, slot 1.5 in long x 0.013 in wide | | |
| Cooled with kiln (1500°C, 1 hr) | 1 | +0.002 |
| Refired and quenched (1500°C, 1 hr) | 1 | −0.0012 |
| Glazed and cooled with kiln (1500°C, 1 hr) | 1 | −0.001 |
| Glazed and quenched (1500°C, 1 hr) | 1 | −0.004 |
| Glazed and quenched (1400°C, 1 hr) | 1 | −0.004 |
| Glazed and quenched (1300°C, 1 hr) | 1 | −0.004 |
| L-2 Glaze | | |
| Glazed and cooled with kiln (1500°C, 1 hr) | 2 | −0.0005 |
| Glazed and quenched (1500°C, 1 hr) | 2 | −0.006 |
| S-1 Glaze | | |
| Glazed and cooled with kiln (1500°C, 1 hr) | 2 | +0.0004 |
| Glazed and quenched (1500°C, 1 hr) | 2 | −0.0014 |

Rods that were glazed with the L-2 glaze and slowly cooled have low compressive surface forces. Tensile forces were expected because of the high thermal expansion coefficient of the glaze. Apparently, vaporization or reaction with the body changed the composition of the glaze so that it had a slightly lower expansion coefficient than the body. Quenching induced compressive forces in the glaze that were even greater than those observed for the regular glaze.

Rods glazed with the S-1 glaze and slowly cooled have tensile stresses in the surface, as expected based on the thermal expansion coefficients. Quenching of the glazed rods resulted in compressive stresses.

## Flexural strength

Flexural strength data for glazed specimens quenched from 1500°C are presented in Table 2.5 and are compared with the strengths of refired controls. The strength increases resulting from "thermal conditioning" and glazing separately add up to 32,700 psi whereas the increase observed for glazing and quenching was 38,300 psi. The increase for glazing and quenching is 5600 psi greater than the other combined increases. The stresses induced in the glaze by quenching may be a more important strengthening factor than is indicated by this difference for the following reasons:

1. Average strengths as high as 99,000 psi for glazed and quenched specimens have been observed.
2. The glazed layer has a low thermal conductivity so that the cooling rate of the alumina surface in the glazed rods must be lower than that of the unglazed rods for similar thermal treatments. Therefore the quenching effect in the alumina of the glazed rods should not contribute as much to the strength of the glazed rods as it does in the bare rods.

Based upon ring test, rod test and flexural strength data, strength increases and stresses cannot be accounted for simply as

Table 2.5.  Flexural Strength of 96% Alumina, Glazed and Quenched (Rods 0.125 in dia, 1500°C, 1 hr)

| Treatment | No. Specimens | Average Strength[*] (psi) | Strength Increase (psi) |
|---|---|---|---|
| Refired and slowly cooled, controls | 5 | 54,600 | |
| Refired and quenched in forced air | 5 | 71,100 | +16,500 |
| Glazed with regular glaze and slowly cooled | 5 | 70,800 | +16,200 |
| Glazed with regular glaze and quenched in forced air | 5 | 92,900 | +38,300 |
| Glazed with L-2 glaze and slowly cooled | 5 | 70,000 | +15,400 |
| Glazed with L-2 glaze and quenched in forced air | 5 | 96,800 | +42,200 |
| Glazed with S-1 glaze and slowly cooled | 5 | 42,300 | -12,300 |
| Glazed with S-1 glaze and quenched in forced air | 5 | 86,500 | +31,900 |

[*]Four point loading on a two-inch span.

the combination of effects of quenching in forced air and glazing
with slow cooling.  Added strength and residual stress may result
from contraction of the hot interior after the glazed surface has
been cooled and becomes rigid as described earlier.

Results for the L-2 glaze are also presented in Table 2.5.  For
glazed and quenched specimens an average flexural strength of
96,800 psi was observed.  It is evident that the strengthening
effect does not necessarily require a glaze with a lower thermal
expansion coefficient than the body.  The higher expansion coefficient of the L-2 glaze was confirmed by slight crazing observed
when thick layers were applied on flat surfaces of specimens that
were slowly cooled.  In thinner layers more reaction occurs and
the expansion coefficient tends toward a lower value than that of
the body.

Results for the S-1 glaze show conclusively the beneficial
effect of quenching the glaze.  In this case, slow cooling leads
to tensile stresses in the surface and a decrease in strength to
42,300 psi.  However, glazing and quenching changes the stresses
to compressive and increases the flexural strength to 86,500 psi.
The degree of reaction of the glaze with the body is important in
this case too.  When the specimens, glazed with the S-1 glaze,
were fired at 1400°C for one hour, the glaze crazed on cooling.
Firing at 1500°C for one hour resulted in sufficient reaction of
the glaze with the body so that crazing was avoided.

Specimens, glazed with the regular glaze, were quenched from
various temperatures.  Flexural strengths of the glazed and
quenched specimens increased with increase in the quenching
temperature, as expected (Table 2.6).

Compressive stresses in the glaze are effective in preventing
abrasion flaws from weakening the specimens.  To illustrate this
point some as-received and some glazed and quenched specimens were
abraded and the flexural strengths were measured (Table 2.7).
Strengths of the as-received specimens were decreased slightly by
the abrasion treatments.  Examination of the individual results

Table 2.6. Flexural Strength of 96% Alumina, Glazed with Regular Glaze and Quenched from Various Temperatures into Forced Air (Rods 0.05 in diameter)

| Treatment | Treatment Conditions | | No. Specimens | Flexural Strength Data | |
|---|---|---|---|---|---|
| | Temp. (°C) | Time (hr) | | Average Strength (psi) | Strength Increase (psi) |
| "As-received" controls | | | 19 | 44,800 | |
| Glazed and quenched | 1100 | 1 | 5 | 63,600 | +18,800 |
| Glazed and quenched | 1200 | 1 | 5 | 64,500 | +19,800 |
| Glazed and quenched | 1300 | 1 | 5 | 69,800 | +25,000 |
| Glazed and quenched | 1400 | 1 | 5 | 77,600 | +32,800 |
| Glazed and quenched* | 1500 | 1 | 5 | 81,900[+] | +37,100 |

*Fired in fluorine containing atmosphere.

[+]The highest individual flexural strength was 108,700 psi.

Table 2.7.  Flexural Strength of 96% Alumina, Glazed, Quenched and Abraded

| Treatment | Abrasion Conditions | No. Specimens | Average Flexural Strength psi |
|---|---|---|---|
| **Rods 0.125 in diameter** | | | |
| As received | None | 19 | 49,700 |
| As received | 240-mesh SiC, ball milled for 10 min | 5 | 42,600 |
| As received | 60-mesh alumina, ball milled for 30 min | 5 | 43,000 |
| Refired in atm. cont'g fluorine (1500°C, 2 hr) glazed, refired (1400°C, 1 hr) quenched | 60-mesh alumina, ball milled for 30 min | 19 | 93,700 |
| **Rods 2.15 in diameter** | | | |
| Glazed with regular glaze (1400°C, 1 hr), quenched | None | 5 | 77,600 |
| Glazed with regular glaze (1400°C, 1 hr) quenched | 240-mesh SiC, ball milled for 10 min | 5 | 81,700 |

showed that the strongest specimens had been weakened, leaving the specimens with very uniform but slightly reduced strengths. The strengths of the glazed and quenched specimens remain high after abrasion. Clearly, the observed strengthening is not the result of a flaw-free surface. In fact, the strength may have been increased by abrasion. If a surface contains a flaw, the stresses concentrated at the flaw can be decreased by surrounding the flaw with less severe flaws.

Some of the highest strengths were observed when the specimens were treated at high temperatures in an atmosphere containing fluorine, prior to glazing and quenching. Flexural strengths of specimens prepared by refiring in the decomposition products of $CrF_3 \cdot 3\text{-}1/2\ H_2O$ followed by glazing and quenching are presented in Table 2.8. The highest strength of an individual specimen prepared by this general method was 108,700 psi.

The effect of glaze thickness on the strength of glazed and quenched specimens was investigated with the results shown in Figure 2.1. The flexural strength increases with increasing glaze thickness for thicknesses up to about 0.0015 in. The thickness that can be achieved is limited, for a particular glaze, by the tendency of the glaze to run off at high quenching temperatures. However, it might be advantageous to change the glaze composition to obtain higher viscosity at high temperatures or to lower the quenching temperatures or to lower the quenching temperature and apply thicker layers.

Other quenching media were investigated with the results summarized in Table 2.9. The highest average flexural strength was 110,700 psi which was achieved by quenching from 1500°C into silicone oil (100 cSt).

### 2.1.3  Tensile Strength

As shown in Figure 1.4(c), flexural loads lead to combined stress distributions that are much different from those existing

Table 2.8. Flexural Strength of 96% Alumina; Treated in a Fluorine-Containing Atmosphere Followed by Glazing and Quenching (Rods 0.125 in diameter)

| Treatment | Treatment Conditions | | | Flexural Strength Data | |
|---|---|---|---|---|---|
| | Temp. (°C) | Time (hr) | No. Specimens | Average Strength (psi) | Strength Increase (psi) |
| "As-received" controls | | | 19 | 49,700 | |
| Refired controls | 1500 | 2 | 19 | 61,900 | 12,200 |
| Refired in atm containing fluorine | 1500 | 2 | 19 | 71,400 | 21,700 |
| Refired in atm containing fluorine, refired, quenched | 1500,1500 | 2,1 | 19 | 67,300 | 17,600 |
| Refired in atm containing fluorine, glazed, quenched | 1500,1500 | 2,1 | 19 | 95,600 | 45,900 |

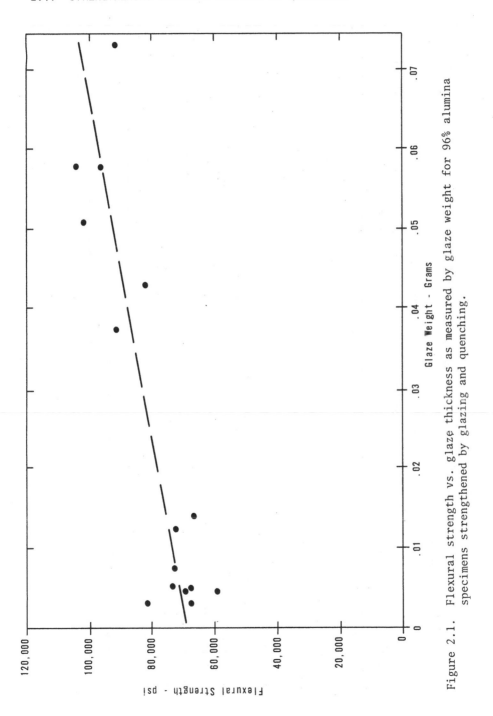

Figure 2.1. Flexural strength vs. glaze thickness as measured by glaze weight for 96% alumina specimens strengthened by glazing and quenching.

Table 2.9.  Flexural Strength of 96% Alumina, Glazed and Quenched in Various Media (Rods Nominally 0.125 in diam)

| Quenching Medium | Quenching Temp. (°C) | No. of Specimens | Average Flexural Strength psi |
|---|---|---|---|
| Silicone oil (100 centistokes) | 1300 | 3 | 78,200 |
| Silicone oil (100 centistokes) | 1400 | 3 | 96,200 |
| Silicone oil (100 centistokes) | 1500 | 3 | 110,700 |
| Silicone oil (100 centistokes) | 1550 | 3 | 105,500[*] |
| Silicone oil (100 centistokes) | 1600 | 3 | 92,900 |
| Motor oil (SAE 30) | 1500 | 3 | 105,000 |
| Forced air | 1500 | 5 | 101,300 |
| Forced helium | 1500 | 5 | 104,000 |
| Forced helium | 1550 | 5 | 108,200 |
| Forced helium | 1600 | 5 | 76,700[+] |
| Forced $CO_2$ | 1500 | 5 | 100,800 |

[*] One result omitted from average on basis of visible flaws.

[+] Glaze crystallized and in poor condition because of high firing temperature.

under tensile loads.  In some cases, depending on the residual stress distribution, the maximum tensile stress under the tensile load may be much greater than that under the equivalent flexural load (by equivalent load we mean here the loads that would lead to equivalent maximum tensile stresses when applied to tensile and flexural specimens that do not contain residual stresses). Therefore it is essential to show that specimens with compressive surface stresses are stronger when loaded in uniform tension as well as when loaded in flexure.

The tensile strength of quenched specimens was measured by the thermal contraction loading method.  One problem is to minimize the mechanical forces exerted by the quenching medium on the specimen so that it will not be deformed while it is in the plastic condition.  Two methods of quenching were used to minimize these forces.  In one case, the necked-down specimens were reheated individually to 1500°C and then cooled by removing them from the furnace into still air.  In the other case, the specimens were lowered gently into viscous silicone oil (12,500 cSt).  In both cases, the specimens appeared to be straight.  The viscous silicone oil was used because some failures caused by thermal shock were observed when less viscous oil was used.

A second difficulty is that the center of the necked-down test section cools more rapidly than the portion nearer the larger ends, because this portion is heated by conduction from the ends.  Therefore, one would expect the strength to vary over the length of the test section.  In an attempt to avoid this problem, "arc"-type specimens were used in some cases.  These specimens were ground by simply lowering the grinding wheel into the surface of the specimen to cut an arc with a radius similar to that of the wheel. Therefore, these specimens have a minimum diameter at the center and since this minimum-diameter point is far removed from the larger diameter ends, the heat flow from the ends is expected to have less effect.

The tensile strengths are presented in Table 2.10.  The average tensile strength of the specimens quenched in still air is

Table 2.10.  Tensile Strength of 96% Alumina

| Sample No. | As Machined* psi | Refired to 1500°C and Quenched in Still Air* psi | Refired to 1500°C and quenched in silicone oil, viscosity 12,500 centistokes+# psi |
|---|---|---|---|
| 1 | 44,300 | 53,000 | 67,400 |
| 2 | 43,000 | 50,000 | 65,200 |
| 3 | 38,000 | 45,000 | 62,100 |
| 4 | 35,700 | 29,400✝ | 57,600 |
| 5 | 30,100 | - - - | 54,100 |
| Average | 38,200 | 49,500 | 61,300 |

*Long specimens, that is, samples with a test section, from shoulder, approximately 3-1/2 in long.

+"Arc"-type specimens.

#One specimen that broke in upper specimen holder not included in average.

✝Result not included in average.

49,500 psi, substantially greater than the strength of the as-
machined specimens.  The average tensile strength of the specimens
quenched in viscous silicone oil is 61,300 psi.

The results of these experiments show that the tensile
strength of the alumina is improved by quenching and the previously
demonstrated improvements in flexural strength are not simply the
result of the distribution of stresses in flexural specimens.

### 2.1.4  Effect of Quenching Temperature on Surface
### Forces and Flexural Strength

Conventional logic suggests that the compressive surface
forces should increase with increasing quenching temperature and
that the flexural strength should increase similarly up to the
point at which thermal shock damage occurs and the strength falls
off substantially.  The increase in residual surface forces, as
indicated by rod tests and accompanying increases in flexural
strength for specimens quenched from various temperatures into
forced air and silicone oil are shown in Figures 2.2 and 2.3.
The increases in strength parallel the increase in residual surface
force, as expected.

Further experiments show that this approach is not sufficient,
however.  This was demonstrated in a striking fashion when specimens quenched from 1450 and 1500°C into linseed oil failed by
thermal shock whereas specimens quenched from 1550°C had an average
flexural strength of 104,700 psi.  This variation in susceptibility
to thermal shock seems to depend on the plasticity of the material
at high temperature.  This phenomenon was investigated in detail
by Gebauer and Hasselman (1971) and Gebauer, Krohn, and Hasselman (1972) for an alumino-silicate ceramic.  They found a well
defined region of intermediate quenching temperatues in which
the material was susceptible to damage and, at higher temperatures,
strengthening was observed as a result of quenching (Figure 2.4).

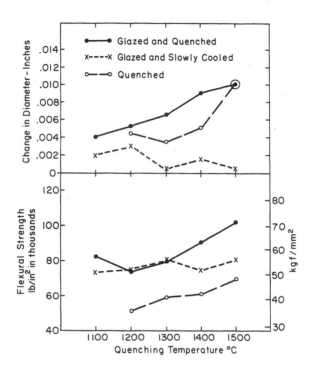

Figure 2.2.   Correlation of compressive stress as measured by
              the rod test and flexural strength of 96% alumina,
              glazed and quenched in forced air.  Reprinted with
              permission of J. Appl. Phys. 42 (10) (1971),
              3685-3692.

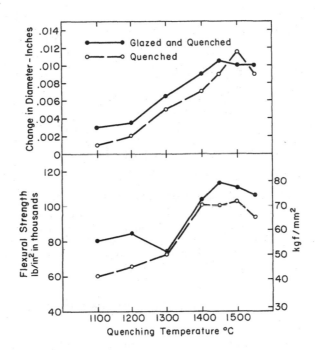

Figure 2.3.   Correlation of compressive stress as measured by the
rod test and flexural strength of 96% alumina, glazed
and quenched in silicone oil (100 cSt).  Reprinted
with permission of J. Appl. Phys. 42 (10) (1971),
3685-3692.

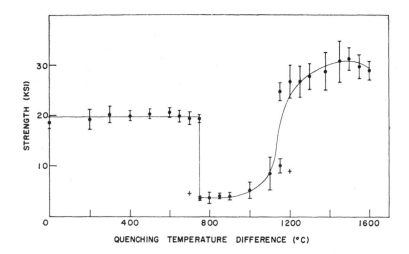

Figure 2.4.    Room temperature strength of aluminosilicate rods
               subjected to thermal shock by quenching in silicone
               oil.  Reprinted with permission of J. Am. Ceram.
               Soc. 54(9) (1971) 468-469.

## 2.1.5  Flexural Strength Distributions

As noted in Section 1.1.7 ceramic materials have acquired a
reputation for unreliability making the distribution of the
individual strength values a subject of considerable interest.
Distribution curves for quenched and glazed and quenched rods are
given in Figure 2.5.  Each group consisted of 19 specimens.  The
average flexural strength of the quenched specimens was 100,600
psi with a coefficient of variation of 2.9%.  The minimum strength
was 95,900 psi.  These results show that the scatter of the
individual strengths of ceramics can be low.  The observations
are important because taking the scatter into account during design
of ceramic parts for structural application is difficult and
expensive and, if ceramics can be shown to be reliable, statistical design procedures can be avoided.

As shown in the data for the 0.137 in diam rods in Table 2.1,
the strength of 96% alumina is consistently increased by refiring
alone raising the question whether the mechanism responsible for

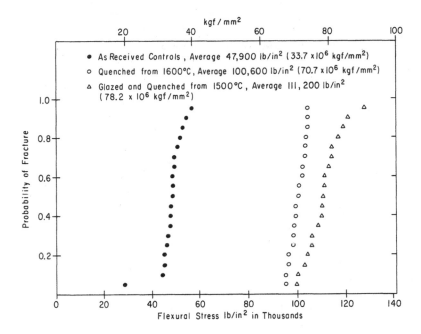

Figure 2.5.   Distributions of flexural strengths of 96% alumina
rods, glazed and quenched in silicone oil (100 cSt).

this increase can in some way be responsible for the larger increases observed for the quenched specimens.  Proposed mechanisms
include flaw healing, rounding of crack tips and relaxation of
localized residual stresses tending to wedge open the flaws.  Distribution curves for as received controls and refired controls are
given in Figure 2.6 and show that the strengths of about half of
the refired specimens fall above the range for the as received
controls.  One possible explanation of this is that there is at
least one type of flaw that is not affected by refiring and con-
tinues to cause failure at the same stress as in the as received
controls.  If this explanation of the strength increase of the
refired specimens is correct, it is apparent that enhancement of
this mechanism cannot yield distribution curves like those in
Figure 2.5.  Therefore, some other mechanism such as one involving
compressive surface stresses must be responsible.

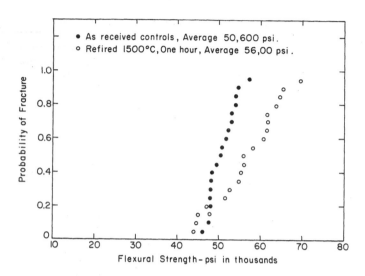

Figure 2.6.    Distributions of flexural strengths of 96% alumina
before and after refiring at 1500°C for one hour.

### 2.1.6  Elevated Temperature Flexural Strength

The elevated temperature strength of alumina ceramics has
been studied by numerous investigators.  Among the recent studies
are those of Davidge and Tappin (1970) who reported strength vs
temperature data for a "pure" (99.7%) coarse grained alumina and
a "debased" (95%) alumina, Crouch and Jolliffe (1970) who measured
the effect of stress rate and temperature on the strength of alumina
refractories, and Congleton, Petch, and Shiels (1969) who measured
the fracture energy at various temperatures.  The strength increases
to a maximum and then decreases sharply.

The residual stress profiles in quenched 96% alumina were
calculated by Buessem and Gruver (1972).  Their work will be
reviewed later.  In the cases they considered, the plastic strains
at the surface and in the interior were of the order of -.001 and
+.001, respectively.  These strains are induced in about two
seconds during quenching from 1500 or 1600°C.  The strength of

96% alumina decreases rapidly with increasing temperature above
800°C and is less than half its maximum strength at 1000°C.  Thus,
a temperature range of practical interest for relief of these
stresses is from about 800 to 1200°C.  If one assumes that the
stresses acting to reduce these residual strains are about 50,000
psi, which is reasonable in view of the magnitude of the observed
strengthening, that a reduction of 10% in the residual strain
would be detectable as a result of its effect on the strength, and
that the creep rate at 1200°C is $10^{-9}$ $sec^{-1}$ $psi^{-1}$, the time required
to produce a detectable reduction in strength is about 0.8 seconds.
At 1000°C the creep rate is about a factor of 20 lower so that the
estimated time is about 16 seconds.  Thus, it is not obvious that
the quenched specimens will retain their improved strength at
elevated temperatures.

The 96% alumina rods were strengthened by quenching from 1500
or 1550°C into silicone oil with a viscosity of 100 cSt (Kirchner
and Gruver, 1974).  In other cases, rods were glazed and quenched
in a similar manner or simply refired to 1500°C and slowly cooled.
The flexural strengths of the 96% alumina rods were measured by
four point loading on a two inch span using a loading rate that
was quite low so that several minutes were required before the
specimens fractured.

The flexural strength vs temperature data for 96% alumina rods
in the as-received condition and with the various treatments are
presented in Figure 2.7.  The flexural strengths of the as-received
rods increased gradually with increasing temperature reaching a
peak of 58,000 psi at about 800°C and dropping off sharply at
higher temperatures.  This peak was observed previously by Crouch
and Jolliffe (1970) who found that, in an 85% alumina refractory,
the magnitude of the peak is dependent on the loading rate and by
Davidge and Tappin (1970) who found that, in a 95% alumina, the
temperature of maximum strength increased with increasing loading
rate.  In addition, these investigators found that the increase in
strength coincided with the onset of non-linearity in the load

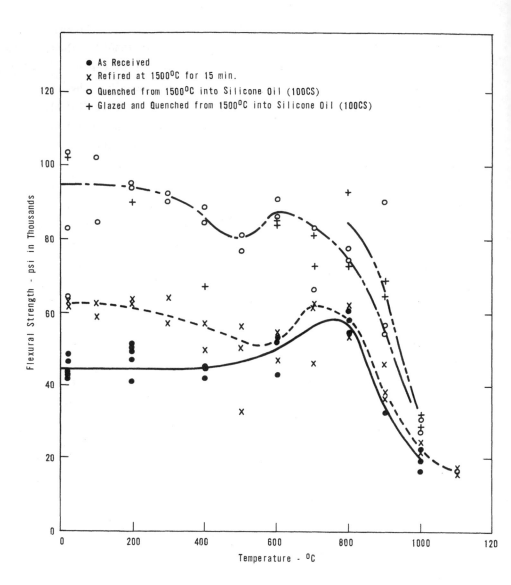

Figure 2.7.   Flexural strength vs. temperature of 96% alumina
with various treatments.  Reprinted with permission
of Mater. Sci. Eng. 13 (1974), 63-69.

deflection curves so that the mechanism of strength increase was interpreted as a dynamic effect involving plastic flow in the glassy phase. Subsequently, a similar strength peak, broader and extending to higher temperatures, was observed for hot pressed alumina which does not have a glassy intergranular phase (Section 2.2.2). Congleton, Petch, and Shiels (1969) have shown that the fracture energy of alumina increases in the temperature range 400-500°C. Therefore, the observed effect may depend on the properties of the alumina as well as the intergranular phase, if present.

The average flexural strength of the rods refired at 1500°C was substantially higher at room temperature than that of the as-received rods. This increase was accompanied by increased scatter in the individual values. Comparison of the distributions of the individual strength values indicates that the strengths of some of the individual specimens are increased by refiring and others are not affected (Figure 2.6). This observation suggests that this increase may be attributable to healing or relief of localized residual stresses of at least one type of flaw population. The strength decreased slowly with increasing temperature. This decrease may be attributable to stress corrosion. After passing through a minimum at 500°C the strength increased, passed through a maximum at 700-800°C and then decline sharply.

At room temperature the average flexural strength of the quenched alumina rods was about twice that of the as-received rods. The strength decreased, passed through a minimum at 500°C, increased slightly at 600°C and then declined sharply. However, at all temperatures the quenched specimens maintained a substantial strength advantage over the unquenched material. In view of the low loading rates, this strength advantage suggests that the residual stresses are relieved slowly so that the improvements in flexural strength could be utilized in at least some elevated temperature applications.

Normally, the glazed and quenched specimens are expected to be slightly stronger than the specimens that are only quenched. In the present case, the room temperature flexural strength was greater than 100,000 psi.  The strengths of specimens tested in the temperature range up to 700°C were very scattered.  At higher temperatures, the strengths were slightly higher than those of the quenched rods, as expected based on the room temperature data. Too few specimens were available to recheck the lower temperature data.

The strengths of larger groups of as received rods were determined at room temperature, 750°C and 950°C.  At 950°C the average strength is lower and the scatter is less than at lower temperatures.  This smaller scatter may be attributable to sub-critical crack growth involving creep or viscous flow.  It is evident that the strengths are not as dependent on the character-istics of individual flaws as is the case at lower temperatures.

Based upon these results, it seems most likely that use of quenched 96% alumina in structural applications would be limited to temperatures under about 800°C.  To determine the effect of longer periods of time at elevated temperature, specimens were held in the test fixture under a slight load at 850 or 900°C for four hours.  Then, the short time strength was measured at that temperature.  The observed strengths were 71,000 at 850°C and 57,000 psi at 900°C, comparing quite well with the strengths expected from Figure 2.7.  Apparently, the residual stresses were not relieved significantly in four hours at these temperatures.

It would be highly desirable to have strength data for quenched specimens exposed to elevated temperatures for much longer periods of time.  However, practical applications may exist for a material with improved strength, even if the duration of the improvement is limited to four hours at 850 or 900°C.

### 2.1.7  Impact Resistance

The improved strength achieved by the use of compressive surface stresses was accompanied by improved impact resistance

(Kirchner and Gruver, 1974). 96% alumina rods were treated by
quenching and by glazing and quenching from 1500°C into silicone
oil (100 cSt). Impact resistances were measured by the drop weight
method. The average values were 0.113 in-lb for as-received,
0.277 in-lb for quenched, and 0.328 in-lb for glazed and quenched
specimens showing that the impact resistance is at least doubled
by quenching and almost tripled by glazing and quenching.

The impact resistance of 96% alumina rods quenched from
1550°C into silicone oil (100 cSt) was measured by the Charpy
method at room temperature and at 100°C intervals over the temperature range from room temperature to 1400°C. The results for rods
of various diameters measured at room temperature (Table 2.11)
show that the energy absorbed during impact is much greater for the
quenched rods than it is for the as-received rods and that the
advantage increases with increasing rod diameter.

The impact resistances of 0.200 in diameter rods tested at
temperatures ranging from room temperature to 1400°C are given in
Figure 2.8. The plotted points are averages of two or three
individual values. The quenched rods maintain their improved
impact resistance up to 1000°C. At higher temperatures the impact
resistance of the quenched rods decreases, perhaps because the
residual stresses in the interior contribute to failure at high
temperatures at which the material itself is substantially weaker
than at room temperature. Similar results, but with increased
scatter, were observed for rods 0.125 in in diameter. The observed
increase in impact resistance is roughly proportional to the
increase in flexural strength. Because the impact resistance is
expected to be proportional to the square of the fracture stress
(Kirchner, Gruver and Sotter, 1975), the increase in impact resistance is less than expected. However, the increase is maintained
to higher temperatures, probably because subcritical crack growth
is limited at high loading rates.

Above 1300°C, the impact resistance of the as-received rods
increased substantially. This increase may occur because of a

Table 2.11.  Impact Resistance of 96% Alumina Rods at Room
Temperature

| Nominal Rod Diameter-in | 0.125 | 0.200 | 0.300 |
|---|---|---|---|
| | Impact Resistance in lb | | |
| As received | 0.23 | 0.72 | 1.72 |
| Quenched from 1550°C | 0.41 | 1.44 | 4.60 |

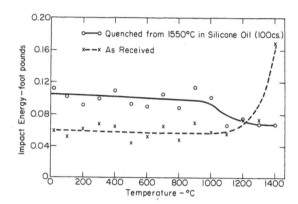

Figure 2.8.  Impact resistance vs. test temperatures for 96% alumina
strengthened by quenching.  Reprinted with permission
of  Mater. Sci. Eng. 13 (1974), 63-69.

transition to less brittle fracture at high temperatues.  Similar
increases are known for metals, glasses and vitreous porcelains.
Kingery and Pappis (1956) looked for such an increase in a pure
alumina body but did not find it in the temperature range from
room temperature to 1600°C.  The increase in impact energy above
1300°C is attributed to a transition to less brittle fracture
involving a decrease in viscosity of the vitreous intergranular
phase.

It seems possible that the peak in flexural strength observed
for the less pure alumina in the temperature range 750-1000°C and

the increase in impact resistance above 1300°C result from the same mechanism. Davidge and Tappin (1970) observed that the temperature of the maximum strength increased about 75°C to 930°C with a 100-fold increase in loading rate so that the specimens fractured in 0.3 sec instead of 30 sec. The time required to load a specimen to the fracture stress during an impact test can be estimated as the time required for the pendulum traveling at its maximum velocity (velocity at impact) to cover a distance equal to the deflection of the rod when it reaches the deflection necessary to cause fracture under a static load. In the present case this requires about $10^{-4}$ sec. Based on the results of Davidge and Tappin a 3000-fold further increase in the loading rate would be expected to cause a substantial increase in the temperature of the maximum strength, assuming that the properties of the body used in this investigation and that of Davidge and Tappin are similar.

Some attempts have been made to estimate this increase in the temperature of the maximum strength. Based on a linear extrapolation, the estimated temperature is much too high. On the other hand, an extrapolation using 1/T vs log 1/t (T is the temperature of the maximum strength and t is the loading time) based on the temperature dependence of viscosity, yielded an estimate somewhat too low. Therefore, it seems likely that the same mechanism is responsible for both phenomena.

### 2.1.8 Thermal Shock Resistance

The thermal shock resistance was determined by requenching the treated rods from an oven into water. The remaining flexural strength after one thermal shock cycle was measured and the temperature change ($\Delta T$) between the oven and the water that the specimens withstood without crack initiation and weakening was used as the measure of thermal shock resistance. The results for specimens quenched from 1450°C into silicone oil are presented in Figure 2.9 (Kirchner, Walker, and Platts, 1971). After quenching

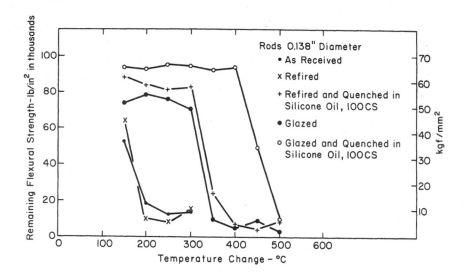

Figure 2.9.   Thermal shock test results for 96% alumina glazed
              and quenched from 1450°C.  Reprinted with permission
              of J. Appl. Phys. 42 (10) (1971), 3685-3692.

from 1450°C, the glazed and quenched specimens survived a ΔT of
400°C without loss of strength.  After quenching from 1550°C,
refired and quenched specimens survived a ΔT of 450°C.  The results
show a substantial improvement over the ΔT of 150°C survived by
the as received rods.

The effect of glaze thickness was investigated and found to
be critical.  Specimens with a thick glaze withstood a ΔT of
325°C but a ΔT of 350°C caused a drastic reduction in strength.
Thinly glazed specimens were weaker to begin with and were
degraded by ΔT > 250°C.

Gupta (1972) and Coppola, Krohn, and Hasselman (1972) have
investigated the thermal shock resistance and strength loss of
alumina ceramics by similar methods.  Gupta found that the temperature change required to cause strength degradation in alumina
specimens of differing average grain size remained the same
(190-200°C) despite substantial variations in strength.  Available

theories usually predict that the required temperature change
increases with increasing strength, other factors remaining equal.
No explanation for this discrepancy is available.  In these investigations the strength degradation was greatest in the strongest
materials, indicating that attempts to improve the thermal shock
resistance by improving the strength of the body are risky because
strength degradation, if it occurs, will be more severe.  If a
crack is initiated, the material may even become weaker than a
material that was initially less strong.  Gebauer, Krohn, and
Hasselman (1972) have shown that strengthening by compressive
surface layers can increase resistance to thermal stress fracture
initiation without increasing the extent of crack propagation.
It is unlikely that the present results for 96% alumina were
determined with sufficient accuracy to be used to evaluate their
theory.

### 2.1.9  Delayed Fracture

As pointed out earlier in Section 1.1.6, alumina ceramics
are susceptible to a stress corrosion effect involving water in
the environment which causes a decrease in fracture stress with
increasing time under load.  Compressive surface stresses reduce
the local stress at a given load leading to the expectation of
improved delayed fracture properties.  96% alumina rods, 0.125 in
diam x 2.5 in long, were loaded in flexure by four-point loading
on a two-inch span in the laboratory atmosphere in which the
humidity was not controlled (Kirchner and Walker, 1971).  The
specimens to be tested in the "as received" condition were proof
tested at 42,000 psi, 30% of the original specimens being removed
to obtain a more uniform distribution.  After proof testing,
groups of five specimens were loaded at each stress level and the
time to failure was measured.  If a specimen survived for four
hours under load, the load was removed and the specimen was set
aside so that the short-time strength could be measured later.

At 36,000 psi, four of the five specimens survived for four
hours. The testing was repeated at intervals of 2,000 psi up to and
including 50,000 psi. All of the specimens loaded at 50,000 psi
broke immediately. The results are plotted in Figure 2.10. The
small numbers over the points at one second indicate the number of
specimens that broke immediately. The numbers at 14,400 seconds
indicate the number of specimens that survived the four-hour test
period. A straight line fits these average values reasonably
well on the semi-log plot.

The fracture surfaces of the as-received specimens broken by
delayed fracture are rather granular in appearance and otherwise
almost featureless reflecting the low stresses at which these
failures occurred.

In order to provide more controlled environmental conditions,
subsequent experiments were performed with the specimens immersed
in distilled water. The results of a comparison of the delayed
fracture properties of 96% alumina in the "as-received" condition
and tested under water and in the laboratory atmosphere are also
presented in Figure 2.10. At short times the alumina withstands
much higher stresses in the laboratory atmosphere than it does
under water. At long times this difference becomes much smaller.

The effect of humidity in decreasing the short-time strength
of 96% alumina is shown by the distribution curves in Figure 2.11
(Kirchner, Gruver, and Walker, 1972) for tests with a stressing
rate of about 50,000 psi per minute. Based on these curves,
the strengths at one second in Figure 2.10 seem reasonable.

## 96% alumina strengthened by quenching

Rods of 96% alumina were quenched from 1600°C into silicone
oil (100 cSt) at room temperature (average flexural strength
101,400 psi). The specimens were loaded in flexure under water
at particular stress levels and the time to fracture was measured.
The resulting delayed fracture curve is presented in Figure 2.10.

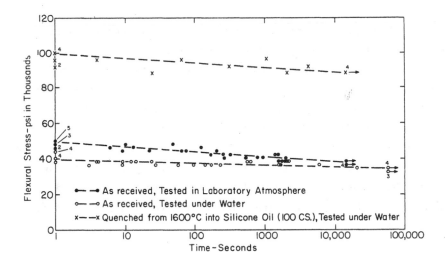

Figure 2.10. Delayed fracture of 96% alumina, as-received and quenched. Reprinted with permission of Mater. Sci. Eng. 8 (1971), 301-309.

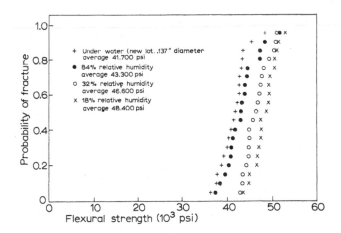

Figure 2.11. Distributions of flexural strengths of 96% alumina tested at various relative humidities (rods 0.125 in diam, four-point loading on a two-inch span). Reprinted with permission of Mater. Sci. Eng. 8 (1971), 301-309.

The quenched specimens maintained their proportionate strength
advantage under the delayed fracture conditions. Elapsed times
many orders of magnitude greater than the test times would be
required to cause failure of the quenched specimens if they were
loaded at the stresses sufficient to cause failure of the as
received specimens in the present tests.

The fracture surfaces of the alumina strengthened by quenching
are very different from those of specimens that were not strengthened. These specimens have well defined fracture mirrors and
hackle, and a large wedge thrown out of the compressively stressed
portion of the specimen.

## 96% alumina strengthened by glazing and quenching

A group of 96% alumina rods were glazed, fired at 1500°C for
one half hour, and quenched in forced air. These rods were proof
tested at 88,000 psi. After proof testing the average flexural
strength was 100,500 psi. The time to failure was measured under
water at stresses ranging from 84,000 to 100,000 psi. Despite
the severe corrosion conditions as a result of the testing under
water, the stresses required to cause delayed fracture were high.
One outstanding specimen survived for more than 100 hours under
water at a stress of 92,000 psi. Compared on the basis of the
ratio of the strength of the treated specimens to the strength of
the controls, the relative strengths remain approximately the
same over the range of testing times.

## Flexural strength of surviving specimens

Specimens that survived the full period of the delayed
fracture test were measured to determine the short-time strength.
The strengths appeared to be increased rather than decreased as
might have been expected assuming that the flaws grow subcritically.
Sedlacek (1968) previously observed a strength increase after
similar stressing. Evans (1974) has shown that in this situation

one should not expect an observable decrease in strength in
materials with very large variations in crack velocity with $K_I$
but he did not consider the possibility of a strengthening effect.

In order to obtain more evidence of the possible strengthening effect of the delayed fracture treatments, several small
groups of as received specimens were stressed in flexure for
similar periods of time. Then, the short-time strengths of those
surviving static fatigue treatment were measured and compared with
appropriate controls. The static fatigue tested specimens were
consistently slightly stronger than the controls confirming the
above observation. It has been argued that the flaws caused by
normal handling in glass heal over a period of time so that the
strength slowly increases during storage. It is reasonable to
hypothesize that such healing could occur when specimens are
stressed to levels insufficient to cause subcritical crack growth
($K_I$ below the static fatigue limit) but further investigation
is needed to determine whether or not such a mechanism applies
in the present case.

The short-term flexural strengths of the four quenched specimens that survived for 14,400 seconds at 88,000 psi were measured.
The strengths of these four specimens were 105,600 psi, 105,200
psi, 103,200 psi and 103,100 psi, averaging 104,300 psi. Because
all of these survivors were stronger than the average before
testing (101,400 psi), the survivors were at least as strong as
the original group, even after allowing for the fact that two
specimens were removed from the group by failure during delayed
fracture testing. At least one of these specimens, when subsequently loaded in the short-time test, failed at an internal flaw.
This flaw consisted of either a pore or low-density region located
well away from the tensile surface. Such a failure at an internal
flaw is further evidence of the effectiveness of compressive
surface stresses in preventing surface flaw failure.

The short-time flexural strengths of the rods, glazed and
quenched in forced air, that subsequently survived the delayed

fracture test, were measured.  The results are presented in Table
2.12.  In every case, the flexural strength is substantially greater
than the proof test level and the stress applied in the delayed
fracture test.  Every specimen that survived the delayed fracture
test was stronger than the original average strength which was
100,500 psi.  These results are substantial evidence that the
stress corrosion does not cause observable deterioration of the
short-time strength of the surface compression strengthened
alumina rods.

## Discussion of results and summary

The delayed fracture curves for the 96% alumina rods tested
in the laboratory atmosphere and under water show that the weakening effect of stress corrosion depends upon the concentration
of water present in the environment.  At short times, the difference in strength between tests in laboratory atmosphere and under
water is about 10,000 psi but at long times this difference
decreases to about 2,000 psi.

Quenched specimens are much stronger than the "as-received"
rods and the absolute decrease in strength during delayed fracture
testing is slightly greater than that of the untreated material.
The similarity in performance indicates that the mechanism of
stress corrosion is the same in both cases.  The main difference
is that for specimens with compressive surface layers, larger
loads are necessary to overcome the compressive surface stresses
to obtain the appropriate level of tensile stress in the surface
so that delayed fracture can occur.

Even though the quenched specimens are susceptible to delayed
fracture, it is advantageous to have the compressive surface layers.
For example, after 10,000 seconds the strength of the quenched
alumina is about 88,000 psi, whereas the strength of the untreated
alumina is only about 34,000 psi.

Table 2.12.  Short-Time Flexural Strength[*] of Glazed and Quenched[+]
Rods Surviving the Delayed Fracture Test (Duration
> 57,900 Seconds)

| Stress Applied in Delayed Fracture Test (psi) | Short-Time Flexural Strength (psi) |
|---|---|
| 84,000 | 103,500 |
|  | 103,000 |
|  | 102,100 |
| 88,000 | 106,200 |
|  | 102,400 |
| 92,000 | 107,700 |
|  | 102,800 |

[*]Four-point loading on a two-inch span.

[+]Glazed with regular glaze, quenched from 1500°C into forced air,
proof tested at 88,000 psi.

The results for the glazed and quenched specimens are quite
similar to those obtained by quenching alone.  In this case a
glassy surface instead of an alumina surface is exposed to stress
corrosion.  The loss of strength of the glazed and unquenched
specimens tested under water is about twice that of the untreated
specimens at equivalent times.  Nevertheless, after 10,000 seconds,
the strength of the treated specimens is about 90,000 psi compared
with about 35,000 psi for the untreated specimens.

The short-time flexural strengths of specimens that survive
the delayed fracture test are as high as or higher than they were
before the delayed fracture test.  This observation casts some
doubt on the conventional description of the mechanism of delayed
fracture, at least for specimens capable of surviving under particular conditions.

Roszhart, Pearson, and Bohn (1971a, 1971b) have shown that
holographic interferometry can be used to obtain highly sensitive
deformation measurements of strained objects.  During investigation
of delayed fracture of glass by this method, the growth of inter-
ference fringes was observed to stop or slow down indicating crack

arrest.  This observation provides additional evidence that the
growth of flaws is not continuous until fracture occurs.

Stress corrosion in the region of the stress concentration
may lead to reduction of the localized stresses if the stress is
not so large that immediate failure occurs.  Grosskreutz (1969,
1970) and Leach (1970) have reported that thin anodized films of
alumina, when exposed to water, have lower elastic modulus, greater
elastic deformation and lower strength than under dry conditions.
Therefore, one mechanism to be considered is that the localized
stress at the most severe flaw is partly relieved by elastic
deformation of the lower modulus alumina surface in contact with
the water or corrosion product.  This localized elastic deformation
would tend to transfer the load to the stiffer crystals in the
surrounding region.

Even though these anodized films are very thin, they are
thick compared with the width of a crack near the crack tip.
Therefore, if the water has a similar effect on the surface of the
bulk alumina, it would be expected to have a substantial effect
on the stress concentration factor.

According to this reasoning, when the load is removed and
the short-time strength is measured, the most severe flaws that
normally would have acted to cause failure at the higher stresses
characteristic of the short-time test are not subjected to stresses
that are quite as high as they would normally be because of the
presence of the lower modulus material.  Therefore, a slightly
higher load is sustained until failure occurs at another flaw that
was originally somewhat less severe.

If this suggested mechanism turns out to be correct it is
necessary to revise our ideas about the effect of moisture on
the short-time strength because it is still necessary to explain
the observed reduction in short-time strength of as-received
rods with increasing humidity.

### 2.1.10   Strength Degradation Caused by Surface Damage

Much effort has been directed toward understanding the surface
damage that occurs during ceramic machining operations.  Hockey
(1971) observed high surface dislocation densities as a result of
room-temperature abrasion of $Al_2O_3$ single crystals, and Gielisse
and Stanislao (1970) found that extensive damage is caused by the
impact of single-point diamond tools.  This surface damage is
expected to cause failure at a strength level determined by the
surface flaw population.  The resistance of glazed and quenched
specimens to strength degradation by abrasion was described in
Section 2.1.2 and Table 2.7.

Both untreated and quenched $Al_2O_3$ rods were scratched by a
diamond point under various loads (Gruver and Kirchner, 1973).
The principal objective was to evaluate the effectiveness of the
compressive surface stresses induced by quenching in reducing the
penetration of surface damage and strength degradation.

96% alumina rods were strengthened by heating individually in
an induction furnace to 1550°C and quenching into silicone oil
(100 cSt).  The diamond points used for scratching were 75° coned
wheel dressers (1/4 carat).  The $Al_2O_3$ rods were held in a Teflon
V-block under the diamond point while a known load was applied
using weights.  Axial and circumferential scratches were formed.
This scratching process was very slow compared with grinding;
about 5 seconds were required to form a 360° circumferential
scratch.

Glass scratched parallel to the applied tensile stress is
stronger than glass with similar scratches perpendicular to the
applied stress.  A perpendicular scratch acts as a notch or
stress concentrator.  However, there is a certain amount of damage
perpendicular to a scratch which can be expected to effect the
strength, so it is of interest to determine the relative effects
of perpendicular and parallel scratches on strength.

Circumferential scratches weakened the 96% $Al_2O_3$ rods more
than the axial scratches as expected based on the experience with

glass.  Both quenched and untreated 96% $Al_2O_3$ rods from a stronger
lot were scratched circumferentially, and the remaining flexural
strength was measured (Figure 2.12).  In addition, untreated specimens dipped in silicone oil were scratched and evaluated to exclude
the possibility that the lubricating effect of the oil reduced the
damage done by the diamond point.  In both absolute and relative
terms the diamond scratches degrade the strength of the quenched
96% $Al_2O_3$ less than that of the untreated material; the strength
of the former is as much as 270% greater than that of the comparable
untreated specimen.  The improvement in performance cannot be
attributed to the lubricating effect of the silicone oil because
the weakening of specimens that were dipped in silicone oil and
then scratched was similar to that of as-received controls.

In the recent investigation of Marshall, Lawn, Kirchner, and
Gruver (1978) in which quenched 96% alumina was damaged in a
controlled manner by the indentation method and the results were
analyzed as described in Section 1.3.4, the relative strength
advantage of the quenched specimens compared with the controls
was less than that indicated in Figure 2.12.  Therefore, it appears
that the relative strength advantage depends on the way the damage
is induced.

### 2.1.11  Stress Profiles

Strengthening of alumina by quenching is analogous to
strengthening of glass by so-called tempering.  Methods have been
developed and extensively applied for calculating stress profiles
in tempered glass (Weymann, 1962).  Buessem and Gruver (1972)
computed residual stress profiles for quenched 96% alumina rods.
Temperature profiles were calculated from experimental heat transfer data and used to calculate the residual strain using experimental creep rate data.

The problems involved in calculations for 96% alumina ceramics
are somewhat simpler than in the case of glass for two reasons

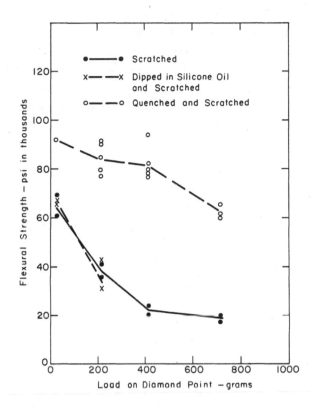

Figure 2.12. Flexural strength vs load on diamond point for scratched 96% alumina. Reprinted with permission of J. Am. Ceram. Soc. 56 (1) (1973), 21-24.

(1) alumina ceramics are less transparent than glass so that heat transfer by thermal radiation within the solid is much less impor- tant, and (2) the crystalline structure of alumina assures that the specific volume of the material depends only on temperature whereas in glass it depends on both temperature and cooling schedule. However, transient creep, that is creep which occurs in the early stages of plastic deformation of alumina and that is characterized by higher creep rates than those observed at longer times, is a complicating factor.

Creep rates were measured at temperatures ranging from 1200 to 1600°C using one eighth inch diameter rods loaded in tension

or in flexure.  The results of these two methods were consistent.
Transient creep was taken into account by assuming that the creep
rate in the early stages was twice that in the later stages.

    The rate of heat transfer from the quenching medium to the
ceramic during quenching was determined by measuring the cooling
time until visible radiation ceased and by measuring the temperature vs. time at the surface and the axis using thermocouples.
The rate of heat transfer was approximately 0.02 cal cm$^{-2}$ sec$^{-1}$ °C$^{-1}$
for quenching in silicone oil (100 cSt).  The temperature profiles
were calculated for several heat transfer rates including values
above and below the experimentally determined rate and for several
different quenching temperatures (Figures 2.13 and 2.14).  The
magnitudes of the computed stresses contain some uncertainty
because of the assumptions involved in the calculations.  Nevertheless, the general relationships of the calculated stresses are

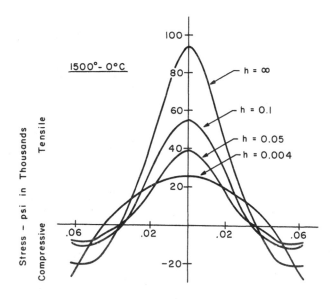

Figure 2.13.  Calculated stress distribution in 96% alumina rods
              quenched from 1500°C, assuming various heat transfer
              coefficients.  Reprinted with permission of J. Am.
              Ceram. Soc. 55 (2) (1972), 101-104.

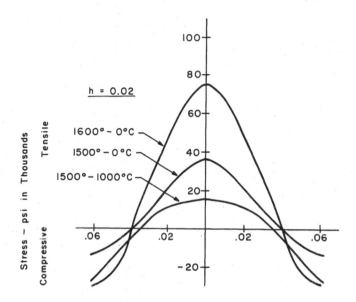

Figure 2.14.   Calculated stress distribution in 96% alumina rods
quenched from various temperatures into a medium
with a heat transfer coefficient of 0.02 cal cm$^{-2}$
s$^{-1}$ $°$C$^{-1}$.  Reprinted with permission of J. Am.
Ceram. Soc. 55 (2) (1972), 101-104.

probably close to those that actually exist.  The calculated
compressive surface stresses range up to about 30,000 psi.  If
it is assumed that the increase in strength cannot be greater
than the residual stress in the surface, the maximum compressive
surface stresses must be about 60,000 psi.  Therefore, these calculations may underestimate the compressive surface stresses.

The tensile stresses in the interior are high enough so that
they may contribute to failure, especially at elevated temperatures at which the material in the interior becomes weaker.  This
problem is discussed in more detail in Section 2.2.2.  A possible
remedy for this problem is suggested by the stress profiles.  They
show that the tensile stresses in the interior can be minimized
by using quenching media having lower rates of heat transfer (h).
Thus, it seems likely that by using moderate quenching temperatures,

together with media having lower rates of heat transfer, it should
be possible to find a set of conditions that minimize this risk of
internal failure.

Semple (1970) determined the residual stresses at the surfaces of alumina bodies subjected to various annealing and quenching treatments, using x-ray diffraction methods.  Unfortunately,
the alumina bodies chosen for investigation were rather weak in
the as-received or as-machined condition so that even air cooled
bars formed surface cracks during cooling, undoubtedly disturbing
the desired residual stress pattern.  Despite this difficulty,
substantial residual compressive stresses were measured in the
surfaces of the quenched specimens.  The maximum value, 27,600 psi,
was observed in the case of 94% alumina quenched from 1650°C.
This result is in reasonable agreement with the calculated results
of Buessem and Gruver (1972) for 96% alumina subjected to a somewhat different quenching treatment.  Semple also developed a method
of calculating the biaxial tensile stresses in the surfaces during
the initial phases of cooling.  This method may be useful for pre-
dicting thermal shock damage.

It has frequently been incorrectly assumed that when the
outermost material is removed from quenched rods by machining,
little stress is present in the new surface.   Stress profiles
were calculated for the case in which the alumina rod is quenched
in forced air and then the material is removed by grinding to
various depths (Kirchner, Buessem, Gruver, Platts, and Walker,
1970) using the principle that the volume averaged stress must be
zero.  The maximum compressive stress in the surface remained
almost constant until the diameter of the rod was reduced by more
than forty percent.  Based upon these calculations, one would
expect the rods to remain strong as material is machined away.
These results are most easily understood by thinking in terms of
the strain gradient.  When part of the compressive surface layer
is removed by grinding, the compressive strain at a particular
point in the interior that is in compression may become greater
although the overall strain in the system has been reduced.  The

tensile strain at a particular point in the interior that is in
tension will become less.  Thus, these changes in strain result in
the shifting of the stress profile.

Alumina rods, 0.125 inches in diameter, were quenched in
forced air or silicone oil.  Then the rods were machined to various
depths and the strengths were measured.  The decrease in strength
after machining away the first thin layer was somewhat greater
than that expected from the stress profile calculations but more
than half of the strengthening remained after removal of several
thousandths of an inch from the diameters and the strengths
remained above control values when the diameters were reduced to
half of their original dimensions.

The stress profiles were calculated for quenched rods subsequently loaded in flexure and are presented in Figure 2.15.  In
the upper figure the residual stress distribution and the stress
distribution due to the applied load are shown separately.  In the
lower figure the stress distribution resulting from combining the
residual and loading stresses is shown.  The figure shows that the
maximum tensile stress increases by only a small amount during
loading.

## 2.1.12  Other Sintered Alumina Bodies

Several other sintered alumina bodies were strengthened by
quenching and glazing and quenching.  At least some strengthening
effect was observed in every case but in some cases the results
were disappointing.  Large grained bodies, sintered to obtain
maximum transparency, are relatively weak in the as-received condition and are adversely affected by reheating in air, perhaps
because of microcracks formed in response to thermal expansion
anisotropy stresses.  Although strengthening was observed, the
strengthened specimens were relatively weak compared with other
strengthened bodies.  Fine grained bodies that were very strong
in the as-received condition were adversely affected by reheating

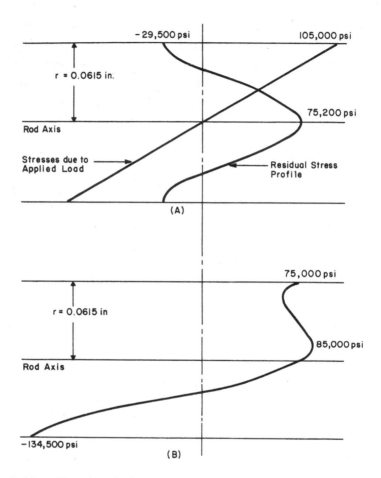

Figure 2.15.    Distribution of stresses at fracture for a 96%
                alumina rod quenched from 1600°C in silicone oil.
                (A) Residual and load stresses.   (B) Combined
                stress distribution.


and little strengthening was observed.  The best results were
obtained with conventional aluminas with a siliceous intergranular
phase and moderate (5 μm) grain size.

## 2.2  STRENGTHENING HOT PRESSED ALUMINA BY QUENCHING

Strengthening of pure hot pressed (H.P.) alumina by quenching differs in some respects from similar treatments for less pure alumina.  One important question is whether or not the presence of a viscous intergranular bonding phawe is essential to formation of compressive surface layers by quenching.  Commercial aluminas usually made by cold pressing and sintering, typically contain silica, magnesia and calcia which, at the reheating temperatures might be expected to be present as a viscous glass.  On the other hand, H.P. alumina bodies are usually 99.5 to 99.9% alumina with 0.1 to 0.25% magnesia added as a grain growth inhibitor.  The amount of second phase in H.P. aluminas is usually very small and is likely to consist mainly of refractory crystals such as spinel.  Thus, if compressive surface layers are formed on H.P. alumina by quenching, the presence of a glassy intergranular phase is not necessary to form compressive surface layers by quenching.

Another important difference between H.P. alumina and the less pure commercial alumina bodies is that the H.P. alumina is stronger.  Therefore, other things being equal, it should be feasible to quench the H.P. alumina more severely and possibly to obtain even greater improvements in strength.

The powders used to form the H.P. alumina were 0.3 μm alumina[*], 0.3 μm deagglomerated alumina[+], and 0.6 μm dry-ball-milled alumina[#] containing MgO (Kirchner, Gruver, and Walker, 1973).  In most cases MgO was added to the first two powders by dissolving magnesium acetate in methyl alcohol, adding the alumina powder, mixing the materials in a blender, and drying with constant

---

[*]Linde A, Union Carbide Corp., New York, New York.

[+]Type CR, Adolph Meller Co., Providence, Rhode Island.

[#]RC-172 DBM, Reynolds Metals Co., Richmond, Va.

agitation. The dried material was loaded into a die, 2-1/2 in in diameter, and heated under a pressure of 4000 psi. Various time and temperature schedules were used.

The billets were cut into rectangular bars which were ground to form cylindrical rods 0.10 to 0.15 in in diameter. To reduce the variability of the strength measurements, these rods were highly polished, using 220, 320, 400 and 600 grit SiC paper and 15 µm diamond paste on paper. The average grain size of the H.P. alumina is approximately one micrometer and it has relatively clean grain boundaries.

Each rod was cemented to a small piece of refactory. The assembly was inverted in a small susceptor in an induction furnace, heated to the desired temperature, removed from the furnace and thrust into the quenching medium. The time required to transfer the specimen was approximately one second. The treated specimens were evaluated by the methods previously described. In the following sections the flexural strength vs. temperature, resistance to penetration of surface damage, residual stresses, and flaws and other microstructural observations are discussed.

### 2.2.1 Flexural Strength

Quenching from temperatures above 1700°C into 12,500 cSt silicone oil frequently resulted in thermal shock failures. These failures which involved cracks perpendicular to the axes of the rods, evidently were caused by axial stresses. Even in specimens quenched from temperatures under 1700°C, thermal shock cracks were observed in some cases. These were axial cracks which were observed as radial black lines in the fracture surfaces and which sometimes extended from the surface to the axis and the full length of the specimen. The black color results from decomposition of silicone oil in the crack. These cracks apparently heal and do not necessarily cause weakness; even the strongest specimens may contain such a crack. The frequency of thermal shock failures also increases at quenching temperatures under 1500°C.

The tendency toward thermal shock failure increased with decreasing viscosity of the quenching medium, and cracks were observed in many cases when rods were quenched from 1700°C into media with viscosities less than 20 cSt.

Rod tests were used to study the relative compressive surface force obtained by quenching into silicone oil. Increasing compressive surface force was observed with increasing quenching temperature. The diameter of a rod, 0.138 in in diameter and quenched from 1700°C into 100 cSt oil, decreased by 0.011 in after it was slotted, showing the presence of compressive surface forces.

## Effect of quenching conditions

Rods made from the 0.3 μm $Al_2O_3$ powder with 0.25 wt% MgO added were quenched into 12,500 cSt silicone oil at temperatures from 1450°C to 1800°C in 50°C intervals. This highly viscous oil was selected to minimize the risk of thermal shock failure. Strength increased moderately, exhibiting a broad maximum from 1500°C to 1700°C. The average strength of the strongest group of specimens was 127,800 psi, compared with the average strength of as-polished controls which usually ranged from 85,000 to 100,000 psi.

Similar rods were quenched from 1700°C into silicone oils of varying viscosities (Table 2.13). The highest average strength, 177,400 psi, was obtained by quenching into 100 cSt oil. The longest remaining section of one of these specimens measured by three-point loading on a 3/4 in span, yielded a strength of 223,000 psi.

## Effect of starting materials

Billets of each of the starting materials were hot pressed in one run at 1425°C for 2 hours at 4000 psi. The polished rods were quenched from 1700°C into 100 cSt silicone oil. The flexural strengths (Table 2.14) were increased by 55 to 70% by quenching. The results for the bodies made from the 0.3 μm and deagglomerated

Table 2.13. Flexural Strengths of H.P. $Al_2O_3$* Quenched from 1700°C into Silicone Oils

| Specimen No. | Flexural Strength⁺ (psi) | | | | | | | |
|---|---|---|---|---|---|---|---|---|
| | 12,500 cSt oil | 1,000 cSt oil | 350 cSt oil | 100 cSt oil | 50 cSt oil | 20 cSt oil | 10 cSt oil | 5 cSt oil |
| 1 | 132,200 | 133,600 | 152,800 | 190,400 | 131,500 | 140,000 | TSF | 120,900 |
| 2 | 131,300 | 130,300 | 149,900 | 181,000 | 129,400 | 109,200 | TSF | 100,500 |
| 3 | 115,900 | 130,000 | 108,900 | 160,700 | 126,200 | TSF# | TSF | TSF |
| Average | 124,500 | 131,300 | 137,200 | 177,400 | 129,400 | 124,600 | | 100,700 |

* Linde A; 0.25 Wt% MgO added.
⁺ Four-point loading on 1-in span.
# TSF = thermal-shock failure.

Table 2.14.   Flexural Strengths of H.P. Aluminas* Quenched from 1700°C into 100 cSt Silicone Oil

| Specimen No. | Flexural Strength+ (psi) | | | | | |
| | 0.3 μm $Al_2O_3$+0.25 wt% MgO (99.85% of theor. density) | | 0.3 μm deagglomerated $Al_2O_3$+0.25 wt% MgO (99.52% of theor. density) | | 0.6 μm $Al_2O_3$# (99.35% of theor. density) | |
| | As polished | Quenched | As polished | Quenched | As polished | Quenched |
|---|---|---|---|---|---|---|
| 1 | 92,800 | 136,400 | 84,700 | 155,900 | 78,600 | 131,800 |
| 2 | 88,700 | 132,000 | 83,500 | 139,600 | 76,700 | 125,300 |
| 3 | 75,000 | 129,500 | 83,500 | 121,500 | 74,700 | |
| 4 | | 63,400† | | 120,400 | 73,200 | |
| Average | 85,500 | 132,600 | 83,900 | 134,300 | 75,800 | 128,500 |

*No open porosity detected.

+Four point loading on 1 in span.

#As-received powder contained 0.1 wt% MgO.

†Omitted from average on basis of axial thermal shock crack.

powders were very similar. The density of the body made from the
coarser powder was lower, leading to lower strength, but despite
the porosity, quenching improved the strength.

## Effect of grain size

Bodies of various grain sizes were prepared from the 0.3 μm
powder, with 0.25% MgO added, by varying the hot pressing time and
temperature. The grain size dependence of the strengths of rods
quenched from 1550°C (Figure 2.16) was similar to that of the
as-polished material in this restricted grain size range. This
observation suggests that the mechanism of fracture and the actual
stress at the surface when fracture occurs remain the same, but
because the compressive surface stresses allow larger loads to be
carried before fracture occurs, the nominal strength is increased.
The limited data for rods quenched from 1700°C indicate a similar
grain size dependence of strength.

Because MgO was so effective in preventing grain growth, the
MgO addition was omitted in some cases to permit the growth of
larger grains. When coarse grained $Al_2O_3$ (55 μm) was quenched,
the strength remained essentially unchanged. The coarse grained
alumina bodies may contain localized cracks caused by the thermal
expansion anisotropy of the individual grains. These localized
cracks may weaken the body and because failures continue to originate at these cracks, quenching may, therefore, be ineffective
in improving the strength.

## 2.2.2 Elevated Temperature Flexural Strength

The elevated temperature flexural strength data for H.P-
alumina in the polished condition and after quenching from 1550°C
are presented in Figure 2.17. The strength of the polished specimens decreases with increasing temperature, passes through a
minimum at 400°C, increases to a broad peak between 600 and 1200°C
and then decreases sharply. The temperature dependence of the

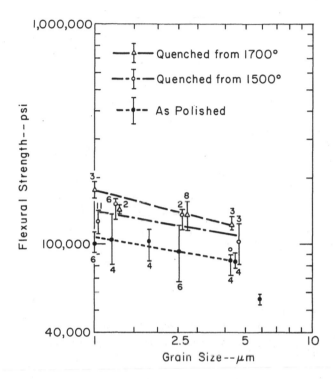

Figure 2.16.    Flexural strength vs. grain size for hot pressed alumina quenched in silicone oil (100 cSt). Reprinted with permission of J. Am. Ceram. Soc. 56 (1) (1973), 17-21.

strength is quite similar to that of the 96% alumina (Figure 2.7) except that the strength values are higher and the decrease in strength at high temperatures occurs at higher temperatures. The similarity, in terms of the peak in the strength at elevated temperatures raises the question whether this peak coincides with the onset of non-linearity in the load deflection curves as found for 95% $Al_2O_3$ by Davidge and Tappin (1970). The answer to this question is not available but it is interesting that, in the present case, very broad peaks are observed compared with those found in other investigations.

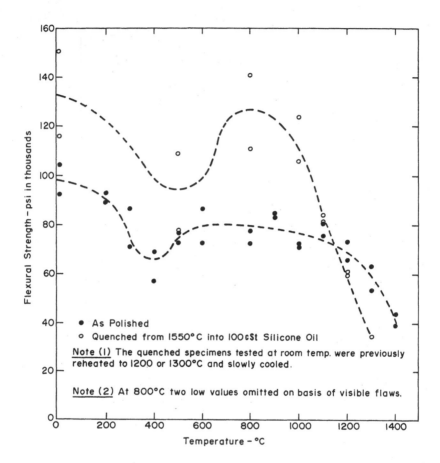

Figure 2.17.   Flexural strength vs. testing temperature for hot
              pressed alumina strengthened by quenching (four-
              point loading on a one-inch span).  Reprinted with
              permission of Mater. Sci. Eng. 13 (1974), 63-69.

The room temperature strength of the quenched alumina is
substantially greater than that of the unquenched material.  Although the data are fragmentary, the strength seems to decrease
and then increase again with increasing temperature in a manner
similar to that of the as polished rods.  Above 1000°C the strength
decreases rapidly so that above about 1150°C the measured strengths
of the as polished alumina are above those of the quenched alumina.

At temperatures of 1100°C and above the scatter of the strengths
is very small indicating that the strength is no longer so dependent
on the characteristics of preexisting flaws.

Just above room temperature the flexural strength of the
polished specimens decreases with increasing temperature, perhaps
because of stress corrosion.  Then, the strength increases perhaps
because of desorption of water, relief of local stresses caused
by expansion anisotropy, or reduction of the severity of flaws by
creep.  Above 1200°C the strength decreases rapidly.

The room temperature strength of the H.P. alumina specimens
that were strengthened by quenching from 1550°C into 100 cSt silicone oil was substantially greater than that of the as polished
rods.  Although the data are fragmentary, the strength seems to
decrease and then increase again with increasing temperature in
the manner similar to that of the polished rods.  Above 1000°C
the strength decreases rapidly.

In comparing the present data for H.P. alumina with that previously presented for 96% alumina, it is evident that the H.P.
alumina, in the polished condition, maintains its strength to
higher temperatures than the 96% alumina (1100°C vs. 800°C).  However, in the quenched condition the strength of the hot pressed
rods falls off more rapidly with increasing temperature than is the
case for the polished rods whereas for 96% alumina high strength
is maintained to relatively higher temperatures in the quenched
condition compared with the as received condition.  This observation is evidence that the residual tensile stresses in the H.P.
alumina, strengthened by quenching, are so high that they contribute
to failure at high temperatures.

The effect of the residual tensile stresses in weakening the
specimens at elevated temperatures was even more evident in the
case of specimens quenched from 1700°C.  The average flexural
strength of these specimens, measured at room temperature, was
153,000 psi.  When these specimens were reheated, they fractured
spontaneously.  In one case the temperature at which fracture
occurred was measured and found to be in the range from 1100-1200°C.
This temperature range is the range in which the strength of the
untreated alumina begins to decrease sharply.  Therefore, it

appears that when the material in the interior is weakened because of increasing temperature, the residual tensile stresses are sufficient to fracture the specimen.

The fracture surfaces of these specimens were studied. In some cases the fractures were flat surfaces perpendicular to the axis. These fractures may be caused mainly by axial residual stresses. In other cases the fractures showed curved surfaces inclined to the axis, or curved chips that popped out of the sides of the rods. There is evidence that the circumferential residual stresses were great enough to cause cracks. Several of the specimens showed curved cracks in the interior as illustrated in Figure 2.18. Apparently, these cracks formed to relieve residual tensile stresses in the interior. The crack propagated until this stress was reduced sufficiently and then it stopped. In some cases these cracks extended to the surface. It seems likely that the curved fracture surfaces originated at or were formed by these cracks.

Figure 2.18.  Fracture surface of a severely quenched alumina rod that fractured spontaneously on reheating to 1100-1200°C.

### 2.2.3  Penetration of Surface Damage

H.P. alumina rods were polished and quenched from 1700°C into silicone oil.  The flexural strengths of scratched rods were measured and compared with similarly scratched controls (Figure 2.19).  As in the case of the 96% alumina, the strength of the quenched H.P. alumina decreased less than that of the untreated alumina in both absolute and relative terms.  The quenched alumina is as much as three times stronger than the comparable untreated alumina, after scratching.  The remaining strength of the quenched alumina decreases as the load on the diamond point increases and then increases again.  This variation is not yet understood.

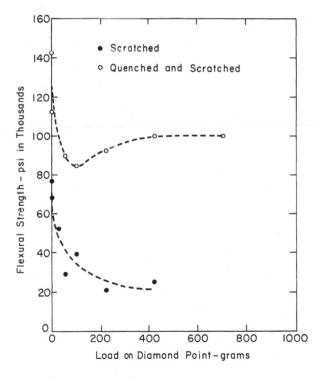

Figure 2.19.  Flexural strength of hot pressed alumina scratched circumferentially with a diamond point.  Reprinted with permission of J. Am. Ceram. Soc. 56 (1) (1973), 21-24.

The fracture surface of strong fine grained alumina (Figure 2.20[A]) is similar to that of glass showing the fracture origin at the surface (at the point of contact of the halves), a mirror region, white patches (flakes caused by crack branching), and hackle (radiating ridges and valleys). If a body that normally is strong is weakened by scratching, the fracture surface is flat and featureless (Figure 2.20[B]), except that these fracture surfaces consistently show evidence of surface damage in the form of "spikes." It is not evident from examination of these fractures how far the damage penetrated during the scratching and what portion of the "spikes" formed during subsequent fracture.

The track formed by the diamond was examined more closely by scanning electron microscopy (Figure 2.21). The scratch is shallow, 30 µm wide, and consists of many individual grooves. The diamond point was examined before and after use; before it was used, the tip was ≃ 10 µm in diameter, and after use it was ≃ 1000 µm in diameter. In addition, many small bumps were formed on the point. Thus, the unexpected shallowness of the scratch can be explained by wear of the diamond point, and the parallel grooves can be attributed to the bumps. Extensive plastic flow of the alumina is evident in Figure 2.21[A]. Gielisse and Stanislao (1970) made similar observations.

To obtain evidence of the depth of penetration of the scratch damage, rods were fractured perpendicular to a scratch, and the cross section was examined. To avoid propagation of the damage, the following procedure was used. An axial scratch was formed on the H.P. alumina rods, which were cut part way through perpendicular to the axis from the side opposite the scratch. The rods were then broken from the side opposite the scratch. The rods were then broken from the cut toward the scratch. Thus, any scratch damage that might propagate was directed toward the scratched surface. The fracture surface of such a rod is shown in Figure 2.21[B]. In the lighter portion of the photograph, the scratch is shown intersecting the edge of the fracture; the darker portion is the fracture surface. The roughly V-shaped features which appear

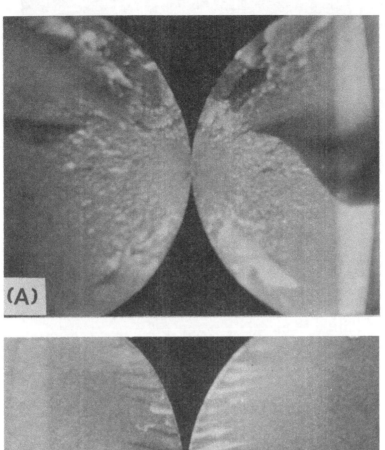

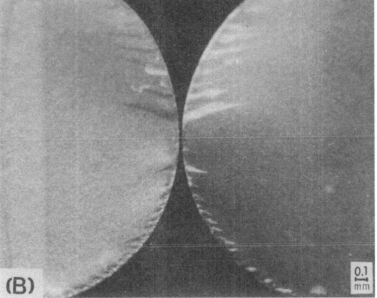

Figure 2.20. Comparison of fracture surfaces of hot pressed alumina rods. (A) As-polished. (B) Scratched. Reprinted with permission of J. Am. Ceram. Soc. 56 (1) (1973), 21-24.

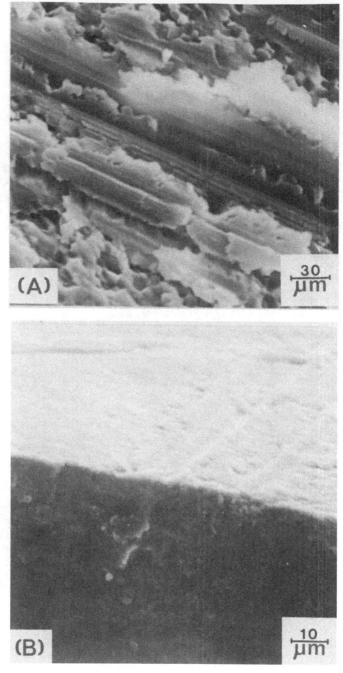

Figure 2.21. Scratch damage in hot pressed alumina. (A) Viewed from
above. (B) Viewed in cross section. Reprinted with
permission of J. Am. Ceram. Soc. 56 (1) (1973), 21-24.

well below the scratch illustrate subsurface damage caused by the diamond point. This damage penetrated ≃ 25 μm below the surface, a distance much greater than the grain size of the material.

The fracture surface of a quenched alumina rod (Figure 2.22 [A]) is quite similar to those of unquenched rods. However, in the fracture surfaces of rods scratched after quenching the mirror broadens along the circumference as the load on the diamond point increases (Figure 2.22[B]) indicating that, as this load increases, more of the scratch acts as a fracture origin.

The strength of scratched, quenched rods remains high enough that mirror boundary formation is observed even in these small specimens. Also, the fact that there is no evidence of the spikes which were observed in the surfaces of the untreated rods is considered to be substantial evidence that the compressive surface layers reduce the penetration of surface damage.

Scratches in the surfaces of quenched rods were examined. The widths and depths of the scratches were approximately the same as for the untreated rods. The most notable difference was the spalling of thin chips from beside the scratch. These chips were up to 30 μm wide. There was a grating sound when the quenched rods were scratched that was not audible for unquenched rods. This noise may have been caused by the chip formation.

In addition to increasing the nominal stress at which surface flaws act to cause failure, the compressive surface layers reduce the penetration of surface damage during diamond scratching. Therefore, scratching reduces the strengths of the quenched rods less than those of untreated rods.

### 2.2.4 Stress Profiles

An estimated stress profile for H.P. alumina quenched from 1700°C into silicone oil was constructed mainly by quantitative analysis of fracture surfaces, especially fracture mirrors (Kirchner and Gruver, 1973). Analysis of fracture surfaces of

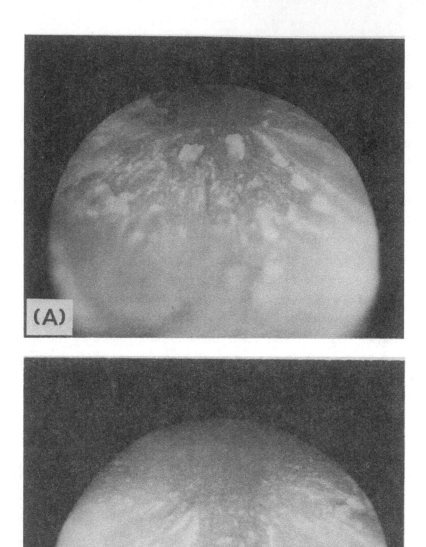

Figure 2.22. Comparison of fracture surfaces of hot pressed alumina.
(A) Quenched. (B) Quenched and scratched. Reprinted
with permission of J. Am. Ceram. Soc. 56 (1) (1973),
21-24

polycrystalline ceramics is handicapped by lack of understanding of the fracture process and variability in the fracture features. In many cases it is difficult to locate fracture origins because the fracture surfaces are flat and featureless.  On the other hand, fractures in glass usually show well-defined and reproducible fracture features including the critical flaw, mirror and mirror boundary, hackle, etc.  Crack branching occurs at or near the hackle boundary.  Using these features, methods of analysis of glass fracture surfaces have been extensively developed.  Much attention has been focused on the study of mirrors.  The theoretical basis for this work was outlined in Section 1.3.3.

Recent improvements in the preparation of alumina ceramics, including the use of hot pressing to obtain dense, fine grained bodies, have made available ceramics that form well defined fracture features.  Therefore, it is now possible to apply the techniques used to analyze the fracture surfaces of glass to the analysis of the fracture surfaces of polycrystalline ceramics.

Mirror radii were measured by optical microscopy using an eyepiece micrometer disc ruled to $5 \times 10^{-5}$ m.  For mirrors formed by fractures originating at surface flaws, the mirror radius was determined by measuring the distance from the fracture origin to the hackle.  For fractures originating internally, the mirror radius was determined by measuring the distance from the hackle on one side to the hackle on the other side of the fracture origin and dividing by two.

The flexural strengths were measured by four point loading on a one inch span at a stressing rate chosen so that the specimens usually fractured in 50 to 100 seconds.  The humidity of the test chamber was controlled at 20% relative humidity for the room temperature tests.

### Fracture stress vs. mirror size

Typical mirrors of fractures originating at the surfaces of
alumina rods are illustrated in Figure 2.23.   Comparing the mirrors
in Figures 2.23[A] and [B] shows the increase in mirror size with
decrease in fracture stress.  In addition, the other fracture features, including hackle and flakes caused by crack branching, become
less pronounced.

The variation of mirror size with fracture stress at normal
loading rates is given in Figure 2.24.  The present results for
H.P. alumina are compared with data, generously provided by R. W.
Rice (1972), from rectangular bars of 94, 96, 98 and 99+% alumina.
The slope of the line based on the present data is 0.46 instead
of 0.5 which one would expect based on the above equations.  These
observations confirm that these equations are applicable to polycrystalline alumina.  A value of A for hot pressed alumina tested
at room temperature is 10.3 $MNm^{-3/2}$ (Kirchner, Gruver, and Sotter,
1974).

There are substantial errors in measuring the mirror radii
because the mirror boundaries are not well defined and the mirror
shape may not be completely symmetrical.  Because there are large
changes in mirror radius with small changes in fracture stress as
indicated by Equation (1.6) the data are useful despite these
large errors.

### Estimating residual stresses based on mirror dimensions

The similarity of mirror size for alumina rods subjected to
very different applied stresses can be observed by comparing
Figure 2.23[B] showing the mirror for an as-polished rod and
Figure 2.23[C] showing the mirror for a rod strengthened by a
compressive surface layer formed by quenching.  The mirror

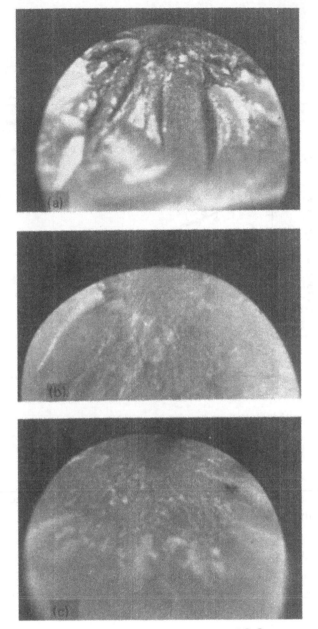

⊢ 100 μm

Figure 2.23.  Comparisons of fracture surfaces of hot pressed alumina.
(A) As-polished, $\sigma_f$ = 806 MNm$^{-2}$.  (B) As-polished, $\sigma_f$ =
551 MNm$^{-2}$.  (C) Quenched from 1700°C in silicone oil,
nominal stress 977 MNm$^{-2}$.  Reprinted with permission of
Phil. Mag. 27(6) (1973), 1433-1446.

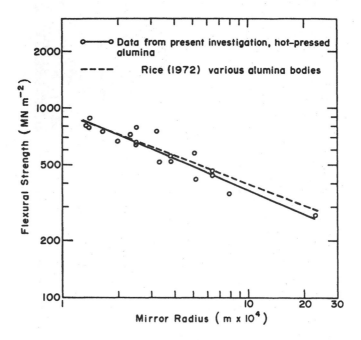

Figure 2.24.   Stress at fracture origin vs. mirror radius for hot
               pressed alumina rods.  Reprinted with permission of
               Phil. Mag. 27 (6) (1973), 1433-1446.

dimensions and flexural strengths of alumina rods quenched from
1550 and 1700°C, for fracture originating at surface flaws, are
given in Figure 2.25.  The difference between the results for the
quenched rods and those for the as-polished rods represents the
residual stress introduced by quenching.  Evaluated at a mirror
radius of $2.5 \times 10^{-4}$ m, the difference between the curves indicates residual stresses of 300 $MNm^{-2}$ obtained by quenching from
1550°C and 400 $MNm^{-2}$ obtained by quenching from 1700°C.  These
residual stress values are in reasonable agreement with the
measured increases in strength.

## Residual stress near the rod axis

H.P. alumina rods that are severely quenched fracture spontaneously when they are reheated to 1100-1200°C.  The mirrors
at these fracture origins were measured and the residual stresses
were estimated.  There is only a small change in the fracture

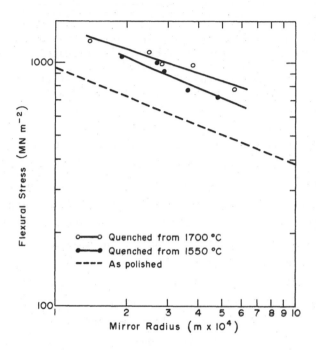

Figure 2.25.   Stress at fracture origin vs. mirror radius for
quenched alumina rods.  Reprinted with permission
of Phil. Mag. 27 (6) (1973), 1433-1446.

stress-mirror radius curves in temperature (Kirchner, Gruver,
and Sotter, 1974).  Assuming that this change with temperature
is not significant, the most frequent stress at or near the rod
axis was 520 $MNm^{-2}$.  Figure 2.18 illustrates one of these mirrors.
Because there is no applied load, the stress at which the spon-
taneous fracture occurs is the strength of the material under
these particular conditions.  This estimated axial stress compares
very well with the flexural strength of the H.P. alumina at 1100-
1200°C.

### Residual stress profile

A residual stress profile for specimens quenched from 1700°C
into silicone oil (100 cSt) was constructed based upon the following information and conditions:

1.  The residual compressive stress at the surface is 400 $MNm^{-2}$
    (from Figure 2.25).
2.  The residual tensile stress at the axis is 520 $MNm^{-2}$, based on
    the mirrors in spontaneous fractures at elevated temperatures.
3.  The volume averaged stress is zero.
4.  The general shape of the stress profile is similar to those
    calculated by Buessem and Gruver (1972) and described in
    Section 2.1.11.

The stress profile is presented in Figure 2.26.  Comparing this
profile with those calculated by Buessem and Gruver (1972) for
96% alumina quenched from 1500 and 1600°C shows reasonable
increases in surface compressive stresses with increasing quenching
temperature.  The residual tensile stress at the axis is less than
might be expected from the calculations but this may result from
the differences in the materials.

It is interesting to estimate the local stress at failure
for quenched specimens failing at internal flaws by combining the
residual stress profile and the local stresses due to the applied
load.  Only axial stresses are accounted for in this estimate.  The
results are plotted in the upper right hand quadrant of Figure
2.26 and show a monotonic decrease in strength with distance from
the rod axis.  The low values occur in the region of greatest slope
of the residual stress profile.

## 2.2.5  Flaws at Fracture Origins

Flaws at fracture origins were located by the methods described in Section 1.2.1 and characterized by optical and scanning
electron microscopy.  For untreated specimens most of the fractures originated at surface flaws but a substantial number of
fractures originated at internal pores and large crystals (Kirchner,
Gruver, and Sotter, 1976).  For example, a surface check or crack in
the fracture surface of an untreated specimen is illustrated in
Figure 2.27.  In this case the specimen fractured at 90,000 psi.

When alumina is strengthened by quenching, the presence of
the compressive surface layer raises the nominal stress at which

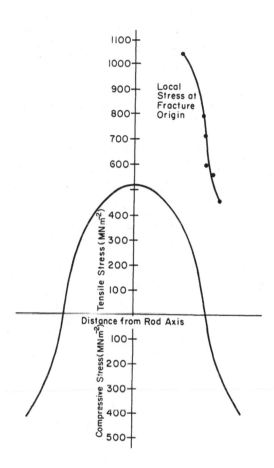

Figure 2.26.   Estimated stress profile for hot pressed alumina
              rods quenched from 1700°C into silicone oil (100
              cSt).  Reprinted with permission of Phil. Mag. 27
              (6) (1973), 1433-1446.

surface flaws act to cause failure.  In the majority of cases
observed thus far the fracture origin is transferred to an internal flaw.  These internal fracture origins may be large isolated
pores, clusters of small pores, large isolated grains, clusters of
large grains, poorly bonded regions, etc.  One such fracture origin
was a large, irregularly shaped pore (Figure 2.28) in a specimen

Based on the mirror radius, the stress acting at the pore was
approximately 830 MNm$^{-2}$. Numerous statements appear in the literature asserting the effectiveness of irregularly shaped pores as
stress concentrators and fracture origins. It should be noted that,
because the stress required to originate fracture at this pore
exceeds the strength of the untreated material, it is unlikely that
this internal pore would have acted as a fracture origin if the
specimen were not strengthened by quenching.

Evidence that transgranular fracture leads to the reflecting
spots is shown in Figure 2.29. Comparison of the scanning electron
micrographs shows the flat areas near the flaw which are not as
frequently observed in the region remote from the flaw.

Fractures originating at groups of coarse grains are of
interest in relation to determination of the variation of strength
with grain size in polycrystalline ceramic bodies. The position
of such a fracture origin in a specimen quenched from 1700°C into
100 cSt oil and having a flexural strength of 1090 MNm$^{-2}$ (158,000
psi) is shown in Figure 2.30[A]. The reflecting spots surrounding
the fracture origin are illustrated in Figure 2.30[B]. Groups of
large grains sometimes form in hot pressed alumina because of a
nonuniform distribution of the grain growth inhibitor or the
presence of localized impurities that compete for the grain
growth inhibitor depleting the grain growth inhibitor available
to the alumina. This group of large grains (Figure 2.30[C]) contains grains ranging in size up to 30 μm. The flexural strength
of H.P. alumina with 30 μm average grain size is approximately
207 MNm$^{-2}$. In the present case the local stress at fracture is
approximately 930 MNm$^{-2}$, based on the mirror diameter. Even
though a body with 30 μm average grain size contains some larger
grains, preventing a direct comparison of the two cases, the
difference in the two stresses is so great that it suggests that
some other factor in addition to grain size is important. This
other factor may be the location of the large grains, that is
whether they are at the surface or in the interior. If this speculation is correct it means that the strength vs. grain size data
that have been measured for alumina ceramics are surface properties.

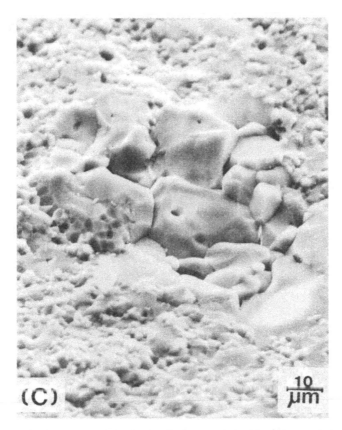

Figure 2.30 (cont.).   (C) Large crystals at fracture origin.  Re-
                      printed with permission of Phil. Mag. 27(6)
                      (1973), 1433-1446.


    The fracture surface and origin in a specimen that failed at
a nominal stress of 223,000 psi are shown in Figure 2.31.  The fracture originated nearer the rod axis than in any other case, the
mirror is small, and the hackle formed relatively sharp ridges and
flat valleys.  The fracture origin appears to be a pore or porous
region associated with grains that are slightly larger than those
in the bulk of the material.  The fracture origin was approximately
25 μm in size, again much larger than the average grain size.

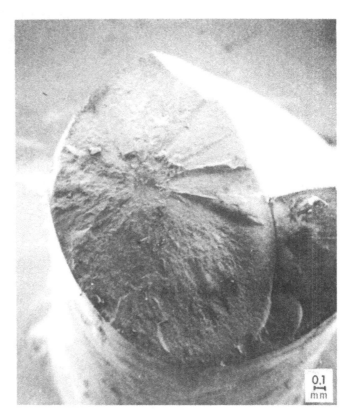

Figure 2.31.  Fracture surface of quenched alumina fractured at a
              nominal stress of 223,000 psi.

2.3 Strengthening Sapphire by Polishing, Reheating and Quenching

   Very high strengths were observed by Morley and Proctor (1962)
and Mallinder and Proctor (1966) for flame polished sapphire rods
tested in flexure.  Chemical polishing in molten borax also resulted
in high strengths (Mallinder and Proctor, 1966).  The effect of
gaseous environments on the fracture behavior of sapphire crystals
was investigated by Wachtman and Maxwell (1954) and Mountvala and
Murray (1964).

   Sapphire rods,* 0.10 in in diameter with the c-axis at a 60°
angle to the rod axis and with surfaces flame polished at the factory
were given the following treatments (Kirchner, Gruver, and Walker,
1969):

---

*Union Carbide Corporation, Crystal Products Div., San Diego, CA.

1.  Fired at 1500°C for one hour and slowly cooled.
2.  Fired at 1500°C for one hour and quenched in forced air.
3.  Glazed with the regular glaze as described in Section 4.2 and
    quenched in forced air, 1500°C, one hour.

Slotted rod tests were used to determine the presence and relative
magnitudes of the compressive forces with the results given in
Table 2.15.  The as-received rods give a false indication of
residual tensile stresses (slot opens).  This change occurs because
of numerous small cracks formed in the surface of the slot.  These
cracks do not close completely so that, effectively, the material
in the surface of the slot occupies larger volume than normal thus
forcing the slot to open.  This surface damage was observed by
optical microscopy.  Reheating, followed by slow cooling or quenching, did not alter the rod test results.  It is likely that the
quenching temperature (1500°C) was too low to induce significant
residual compressive stresses in sapphire.  Glazing and quenching
resulted in substantial compressive surface forces indicated by the
fact that the diameter of the slotted rod decreased by 0.004 in.

Table 2.15.  Rod Test Results for Sapphire (Rods 0.1 in diameter,
             Slot 1.125 x 0.013 in)

| Treatment | Change in Rod Diameter |
| --- | --- |
| As-received control | +0.002 |
| Fired at 1500°C for one hour and slowly cooled | +0.002 |
| Fired at 1500°C for one hour and quenched in forced air | +0.002 |
| Glazed and quenched in forced air, 1500°C, one hour | -0.004 |

The composition profile for the surface of a glazed rod
quenched from 1500°C into forced air was determined by electron
microprobe analysis (Figure 2.32).  As a result of the one hour
hold period at 1500°C before quenching, most of the lead evaporated
from the glaze as described in Section 2.1.2.  Evaporation of boron
and the alkalis would be expected, also.  There is little indication of compound formation at the glaze-crystal interface but
there is some evidence of penetration of silicon, magnesium and
lead into the sapphire.

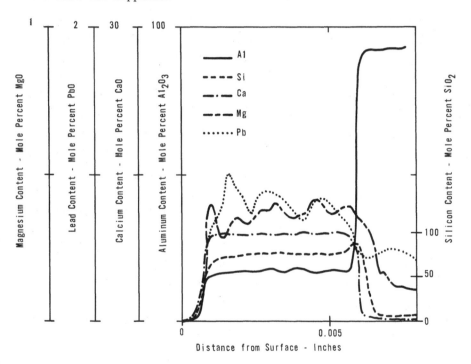

Figure 2.32.  Composition profiles of a sapphire single crystal,
              glazed and quenched (1500°C, one hour).  Reprinted
              with permission of J. Appl. Phys. 40 (9) (1969),
              3445-3452.

### 2.3.1  Flexural Strength

The flexural strengths of sapphire rods, treated by the methods
described above, were measured by three-point loading on a one inch
span (Table 2.16).  Before the strengths were measured, the specimens were oriented by determining the position at which extinction
occurs in transmitted polarized light.  When the orientation of
the crystal is such that the plane determined by the c-axis and the
rod axis is the same as that determined by the rod axis and the
load vectors, the specimen is described as being in the 0° orientation.  Simply refiring followed by slow cooling yields a substantial
strength improvement.  Because flame polishing is known to yield
very high strengths, it is evident that the as-received strength
depends mainly on the damage that has occurred since the rods were
originally flame polished at the factory.  Any observed orientation
dependence of the strength of the as-received rods reflects mainly
directional variations in resistance to surface damage.  The
improvement on refiring may be due to healing of surface damage,
annealing of localized stresses at flaws, or decomposition of
surface hydrates (Mountvala and Murray, 1964).

Quenching the unglazed rods in forced air did not result in
a further improvement in strength; in fact, these rods were not as
strong as those that were simply refired.  This observation is
consistent with the rod test results which showed that quenching
from 1500°C into forced air did not induce significant compressive
surface forces.  The forced air stream may have contained dust
particles or oil droplets which reintroduced some surface damage.

Glazing, followed by slow cooling, and glazing and quenching
in forced air were the most effective treatments.  Because the
glaze has a lower thermal expansion coefficient than the body, the
glaze was under compression after cooling to room temperature.
Quenching the glazed rods yielded an additional increase in strength.

Table 2.16. Flexural Strength of Glazed and Quenched Sapphire (1500°C, One Hour)

| Treatment | No. Specimens | Average Flexural Strength psi | Standard Deviation psi |
|---|---|---|---|
| 0° Orientation | | | |
| As-received | 5 | 88,800 | 25,200 |
| Refired, cooled with kiln | 5 | 190,800 | 38,300 |
| Refired, quenched in forced air | 5 | 116,400 | 44,600 |
| Glazed, cooled with kiln | 4 | 278,000 | 16,900 |
| Glazed, quenched in forced air | 5 | 288,200 (360,000 max) | 45,800 |
| 90° Orientation | | | |
| As-received | 5 | 104,900 | 23,900 |
| Refired, cooled with kiln | 5 | 162,700 | 87,900 |
| Refired, quenched in forced air | 5 | 138,600 | 39,800 |
| Glazed, cooled with kiln | 4 | 301,000 | 44,700 |
| Glazed, quenched in forced air | 5 | 308,000 (349,000 max) | 33,500 |

Treated and control specimens were abraded in a jar mill for 15 minutes in 240-mesh boron carbide (Table 2.17).  The strengths of as-received and glazed and quenched rods were reduced to 47,100 and 217,200 psi, respectively.  In the abraded condition the glazed and quenched rods are more than four times as strong as the as-received and abraded rods.

The glaze has a very low elastic modulus compared with the sapphire rod.  Therefore, as the rod deflects in the flexural-strength test, the changes in stress in the surface layer are small compared with those in the sapphire.  This low elastic modulus effect, plus the presence of the compressive stresses in the glazed layer, are apparently effective in preventing surface flaws caused by abrasion from acting to cause failure.

The possibility of increasing the strength by improving the quality of the crystal surfaces before the treatments was considered.  Flame polishing and chemical polishing were used.  Specimens were polished in 85% phosphoric acid using a reflux apparatus but the rods became elliptical in cross section because the reaction rate varied with crystallographic direction.  Although some improvement in strength was observed, so much material was lost at the higher temperatures by chemical milling (up to 12% reduction in rod diameter) that substantial improvements in strength were unlikely before too much material dissolved.

Borax treatments removed less material; in 20 min at 1000°C the specimen diameter was reduced by only 0.005 in.  Polished rods had average strengths as high as 561,000 psi.  Glazing and quenching resulted in average strengths as high as 333,000 psi.

Flame polishing the specimens previously flame polished at the factory was also effective.  The specimens were rotated in a lathe while they were polished by a gas-oxygen torch which was mechanically traversed along the rods.  Repolishing improved the strength of the unglazed specimens to an average of 421,700 psi (591,000 psi max).  Glazed and quenched specimens had average flexural strengths as high as 340,000 psi, a value higher than those observed for specimens that were not repolished just before glazing.

Table 2.17.  Flexural Strength of Abraded Sapphire (0° Orientation)

| Treatment | No. Specimens | Average Flexural Strength psi | Standard Deviation psi |
|---|---|---|---|
| As-received | 5 | 88,800 | 25,200 |
| As-received and abraded* | 5 | 47,100 | 10,000 |
| Glazed and quenched in forced air, (1500°C, one hour) | 5 | 288,200 | 45,800 |
| Glazed and quenched in forced air, (1500°C, one hour) abraded* | 4 | 217,200 | 26,300 |

*240-mesh boron carbide in a jar mill for 15 minutes.

## 2.3.2 Thermal Shock Resistance

The thermal shock resistance of sapphire was evaluated by heating the specimens to various temperatures in an oven, quenching into water at room temperature and measuring the remaining flexural strength. As-received and glazed and quenched specimens were tested with the results shown in Figure 2.33. As measured by the temperature change required to cause strength degradation (250°C), the sapphire crystals were only slightly more resistant to thermal shock than polycrystalline alumina. When cracks form, they tend to propagate completely through the crystals reducing the strength almost to zero. A temperature change of 500°C was required to degrade the strength of the glazed and quenched specimens (regular glaze, refired at 1500°C and quenched in forced air) indicating a substantial improvement in thermal shock resistance.

## 2.3.3 Analysis of Strength and Thermal Shock Test Data

Wiederhorn (1969) investigated the fracture of sapphire and found substantial variation of fracture energy with crystallographic direction. Subsequently, he (1974) measured the crack velocity vs. $K_I$ in a plane of easy crack propagation ($10\bar{1}2$). A short extrapolation of the data indicates that at a velocity of $10^{-4}$ ms$^{-1}$, $K_I \simeq 1.7$ MNm$^{-3/2}$. If this value is assumed for $K_{IC}$ and the flaws are assumed to be notches that are long compared with their depths the flaw depths (a) can be estimated using Equation 1.5 in which $\frac{Y}{Z} = 1.99$

$$a = \left(\frac{K_{IC}}{1.99\sigma_f}\right)^2 \qquad (2.1)$$

Assuming that localized residual tensile stresses at the flaw are absent so that the strength can be substituted for $\sigma_f$, the estimated flaw depths are $\sim 2$ μm for the as-received sapphire and $\sim 0.06$ μm for flame polished sapphire.

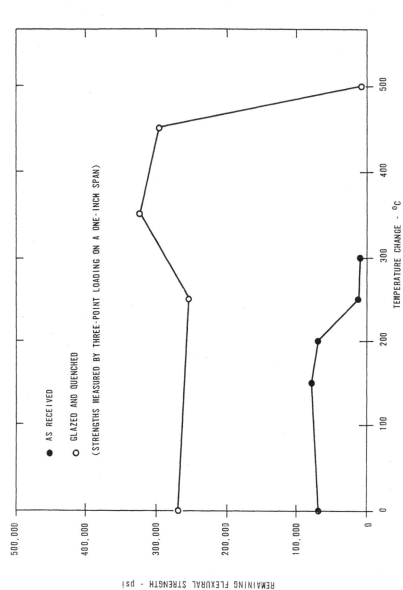

Figure 2.33.   Thermal shock test results for glazed and quenched sapphire (thermal shock treatment was quenching in water at room temperature).   Reprinted with permission of J. Appl. Phys. 40 (9) (1969), 3445-3452.

Calculation of the flaw depths in glazed and glazed and quenched specimens involves added uncertainties.  First, it is uncertain whether the flaw is in the sapphire or the glaze or both. $K_{IC}$ for glaze is about 0.7 MNm$^{-3/2}$ so that estimates based on the assumption that the flaws are in the glaze lead to lower values than otherwise would be the case.  Second, the stress acting to open the crack is the local stress, which when compressive surface stresses are present is the applied tensile stress plus any localized residual tensile stress at the flaw, less the residual compressive stress.  The rod test results indicate that if compressive stresses are present in the sapphire they are not large. Refiring the as-received specimens roughly doubles the strength. This increase may be the result of flaw healing or the result of relaxation of localized stresses (annealing) at the flaw.  In view of the added uncertainties, it doesn't seem worthwhile to calculate flaw depths for these specimens but flaw depths less than about 2 μm should be expected.

The thermal shock resistance results can be analyzed as follows.  If it is assumed that failure occurs during the early stages of cooling before the average temperature of the specimen has changed appreciably from the quenching temperature, a rough estimate of the temperature difference ($\Delta T$) required can be estimated using the equation for an infinite cooling rate

$$\Delta T = \frac{\sigma_f(1-\nu)}{E\alpha} \tag{2.2}$$

in which $\sigma_f$ is the fracture stress, $\nu$ is Poisson's ratio, E is Young's modulus and $\alpha$ is the coefficient of thermal expansion. Using $\sigma_f$ = 88,000 psi and reasonable values for other quantities, $\Delta T$ = 160°C which is in reasonable agreement with the experimental value of 200-250°C.  The principal factor that changes for glazed and quenched specimens is $\sigma_f$ which increases to 288,000 psi (the flexural strength of similarly treated specimens.  Multiplying 160°C by the ratio of the strengths leads to $\Delta T$ = 519°C as an estimate of the temperature difference required to cause failure

of the glazed and quenched sapphire.  Again, the estimate 519°C is
in reasonable agreement with the experimental value of 450-500°C.

    Interpretation of the improvements in strength observed for
sapphire is not as straightforward as it is for polycrystalline
alumina because of the large increases in strength observed as a
result of simply reheating, chemical polishing or flame polishing.
In cases in which the sapphire would not be subjected to surface
damage these other methods, not involving compressive surface
layers, might be satisfactory.  However, most structural applications involve risk of surface damage.  In those cases it would be
desirable to protect the surface by glazing or glazing and
quenching.

## 2.4  Strengthening Other Oxide and Silicate
## Ceramics by Quenching

    The major portion of the effort to develop strengthening
processes was devoted to alumina because of its high strength
which assured that it would be a prime candidate for applications
in which strength is important.  However, in an effort to determine
the range of applicability of the processes, exploratory experiments were done with a wide range of additional oxide and silicate
materials.  A complicating factor is that historically the creep
rates of polycrystalline ceramics were considered to be too low
to allow significant strengthening by quenching.  This situation
led to overemphasis on strengthening by glazing and by glazing and
quenching and failure to quench unglazed specimens in cases in
which it would now seem logical to do so.  In the following sections
strengthening processes for titania, spinel, steatite, forsterite,
zircon porcelain, electrical porcelain and mullite bodies are
described.

## 2.4.1 Titania ($TiO_2$)

Titania bars, prepared from a mixture of 99 3/4% $TiO_2$ powder[*] + 1/4% $WO_3$ by dry pressing and sintering, were quenched from various temperatures into forced air with the results shown in Table 2.18 (Kirchner, Gruver, Platts, and Walker, 1967). Quenching alone increased the strength by about 33%. Glazing followed by slow cooling resulted in a similar increase in strength and when glazing and quenching were combined the strength increased by roughly the sum of the two treatments, 67%.

As will be described in a later section, treatments with fluorine are effective in strengthening titania ceramics. Specimens that were refired at 1400°C in a fluorine containing atmosphere, then glazed and quenched from 1200°C into forced air, were about 70% stronger than as fired controls. Another process for treating with fluorine involves packing the specimens in powders containing fluorides and refiring to temperatures at which the powders volatilize or decompose. Specimens treated by this method were also glazed and quenched and improvements in strength of up to 78% were observed.

## 2.4.2 Spinel ($MgAl_2O_4$)

Hollow cylinders of spinel[+] were quenched from 1500°C into forced air with the results shown in Table 2.19 (Kirchner, Gruver, Platts, Rishel, and Walker, 1968). Quenching increased the strength by 27%. In contrast to alumina and most other materials investigated, refiring followed by slow cooling decreased the strength. Therefore, the entire strength increase in the quenched specimens can be attributed to the quenching process.

---

[*]TAM heavy grade $TiO_2$, TAMCO, Niagara Falls, New York.

[+]DEGUSSA SP-23, DEGUSSA Incorporated, Kearney, New Jersey.

Table 2.18. Flexural Strength of Titania Strengthened by Quenching (3/8 x 1/4 x 3 in Bars)

| Treatment | No. Specimens | Average Flexural Strength* psi | Strength Difference psi |
|---|---|---|---|
| Controls, as fired | 5 | 18,700 | - - - |
| Quenched from 1200°C into forced air | 5 | 24,800[+] | + 6,100[+] |
| Glazed and cooled slowly | 5 | 25,800 | + 7,100 |
| Glazed and quenched from 1200°C into forced air | 5 | 31,200 | +12,500 |
| Controls, as fired | 5 | 17,500 | - - - |
| Glazed and quenched from 1200°C into forced air | 5 | 27,700 | +10,200 |
| Refired in a fluorine containing atm. at 1400°C, glazed and quenched from 1200°C into forced air | 5 | 29,800 | +12,300 |
| Refired packed in 90% $Cr_2O_3$ + 10% $CrF_3 \cdot 1/2\ H_2O \cdot$ powder mixture, glazed and quenched from 1200°C into forced air | 5 | 30,900 (34,100)[#] | +13,400 |
| Refired packed in 90% $SnO_2$ + 10% $AlF_3 \cdot x\ H_2O$ powder mixture, glazed and quenched from 1200°C into forced air | 5 | 31,100 | +13,600 |

*Four point loading on a two inch span.
[+]Three of the five specimens were damage by thermal shock and the results have been omitted from the average.
[#]Average of 4 highest values.

Table 2.19.  Flexural Strength of Spinel* Strengthened by Quenching (Hollow Cylinders 0.238 OD x 0.117 in ID x 2.25 in long)

| Treatment | No. Specimens | Average Flexural Strength[+] psi | Strength Difference psi |
|---|---|---|---|
| Controls, as received | 4 | 15,000 | - - - |
| Refired at 1650°C for one hour | 4 | 13,400 | -1,600 |
| Refired at 1500°C for one hour | 4 | 14,400 | - 600 |
| Refired at 1500°C for one hour and quenched into forced air | 4 | 19,000 | +4,000 |
| Glazed at 1500°C and slowly cooled | 4 | 23,300 | +8,300 |
| Glazed and quenched from 1500°C forced air | 4 | 24,300 | +9,300 |

*DEGUSSA SP-23 Spinel.
[+]Four point loading on a two inch span.

Glazing and slow cooling increased the strength by 55%.  The regular glaze has a thermal expansion coefficient of $\sim 55 \times 10^{-7}\ °C^{-1}$ if little change in glaze composition occurs as a result of evaporation or reaction with the body.  Spinel has a thermal expansion coefficient of $\sim 87 \times 10^{-}\ °C^{-1}$.  Because of the large difference in thermal expansion coefficients, the glaze is expected to be subject to substantial compressive stresses on cooling to room temperature.

Glazing and quenching from 1500°C into forced air increased the strength by 62%.  Thus, only a moderate increase was observed compared with specimens that were glazed and slowly cooled.

One might expect the thermal treatments to apply less well to hollow cylinders compared with solid cylinders because the inner surface is probably in tension in most cases and it is exposed to stress corrosion by water in the atmosphere.  Although strengthening of the spinel was less successful than that of some of the other materials, there is no direct evidence that this was caused by use of hollow cylindrical specimens.

### 2.4.3  Steatite ($MgSiO_3$)

Steatite rods[*], 0.19 in diam, were quenched from various temperatures into silicone oils of various viscosities, with the results given in Table 2.20 (Kirchner, Gruver, and Platts, 1971). In 10 cSt silicone oil the specimens failed by thermal shock but in the more viscous oils substantial strengthening was observed. The best improvement, 73%, was observed in specimens quenched from 1250°C into 100 cSt silicone oil.  Quenching from higher temperatures, 1300 and 1350°C, did not yield further improvements.

Results obtained for additional treatments are included at the bottom of Table 2.20.  These results are not directly comparable to those above because the flexural strength measurements were made

_____

[*] DC-144, DU-CO Ceramics Co., Saxonburg, Pa.

Table 2.20.  Flexural Strength of Steatite* Strengthened by Quenching (Cylindrical Rods, 0.19 in diam)

| Treatment | No. Specimens | Average Flexural Strength[+] psi | Strength Increase psi |
|---|---|---|---|
| Controls, as received | 3 | 28,200 | --- |
| Quenched from 1250°C into 10 cSt silicone oil | 3 | ---# | --- |
| Quenched from 1250°C into 20 cSt silicone oil | 3 | 47,400 | +19,200 |
| Quenched from 1250°C into 50 cSt silicone oil | 3 | 40,100 | +11,900 |
| Quenched from 1250°C into 100 cSt silicone oil | 3 | 48,800 | +20,600 |
| Quenched from 1300°C into 100 cSt silicone oil | 3 | 42,100 | +13,900 |
| Quenched from 1350°C into 100 cSt silicone oil | 3 | 47,600 | +19,400 |
| Refired to 1300°C for one hour | 5 | 37,000[ƒ] | + 8,800 |
| Glazed at 1300°C and cooled with kiln | 5 | 30,500[ƒ] | + 2,300 |
| Glazed and quenched from 1300°C into forced air | 5 | 43,600[ƒ] | +15,400 |

*DC-144, DU-CO Ceramics Co., Saxonburg, Pa.
+Four point loading on a two inch span except as noted.
#Thermal shock failures during quenching.
ƒThree point loading on a one inch span.

by three point loading on a one inch span which normally yields
higher values. Specimens that were refired at 1300°C for one
hour and slowly cooled had an average strength of 37,000 psi.
Perhaps half of the difference is caused by refiring and half by
the difference in testing methods. Specimens glazed at 1300°C
and slowly cooled had an average strength of 30,500 psi which is
lower than might be expected, perhaps because this glaze attacks
the body. Glazed and quenched specimens had an average strength
of 43,600 psi but this result might have been higher if the specimens were quenched into a liquid medium instead of forced air and
if the glaze was more compatible with the body.

Steatite substrates, 0.037 in thick, were quenched from 1250°C
into silicone oils of various viscosities. Specimens quenched
into 5 and 100 cSt oils failed by thermal shock. Specimens quenched
into 12,500 cSt oil had an average flexural strength of 16,400
psi compared with 11,000 psi for similarly tested controls[*]. This
result shows that the quenching process is effective in strengthening thin rectangular plates as well as cylindrical rods.

## 2.4.4 Forsterite ($Mg_2SiO_4$)

Hollow cylinders (0.25 in OD x 0.15 in ID x 3.00 in long) of
forsterite were glazed and quenched from 1300°C into forced air with
the results shown in Table 2.21 (Kirchner, Gruver, Platts, and
Walker, 1967). The strengths of refired controls and specimens
glazed at 1300°C were degraded substantially. Specimens that
were glazed and quenched from 1300°C recovered most of the lost
strength and were only slightly weaker than the as received controls. Therefore, if the as-received controls are used as the
standard for comparison, none of the treatments were effective in
strengthening the forsterite specimens.

---

[*] Three point loading on a one half inch span.

Table 2.21. Flexural Strength of Forsterite[*] Strengthened by Quenching (Hollow Cylinders 0.25 in ID x 0.15 in OD x 3.00 in long)

| Treatment | No. Specimens | Average Flexural Strength[+] psi | Strength Difference psi |
|---|---|---|---|
| Controls, as received | 5 | 19,200 | --- |
| Controls, refired 1300°C for one hour | 5 | 15,500 | -3,700 |
| Glazed at 1300°C and slowly cooled | 5 | 14,700 | -4,500 |
| Glazed and quenched from 1300°C into forced air | 5 | 17,400 | -1,800 |

[*] DC-200 Forsterite, DU-CO Ceramics Co., Saxonburg, Pa.
[+] Four point loading on a two inch span.

## 2.4.5  Zircon Porcelain

Zircon porcelain rods[*], 0.120 in diam, were strengthened by
quenching from 1400°C into forced air with the results shown in
Table 2.22 (Kirchner, Gruver, Platts, Rishel, and Walker, 1968).
Refiring at 1400°C left the strength unchanged.  Quenching from
1400°C into forced air resulted in a small increase in strength
but glazing and slow cooling led to a small decrease.  The greatest
increase (14%) was obtained by glazing and quenching from 1400°C
into forced air.  Thus, the thermal treatments were relatively ineffective in strengthening zircon porcelain.

## 2.4.6  Electrical Porcelain

Electrical porcelain rods were strengthened by quenching into
forced air with the results shown in Table 2.23 (Kirchner, Gruver,
Platts, Rishel, and Walker, 1969).  Two grades were used; one
unglazed[+] and one glazed by the manufacturer[#].  The glazed material
in the as-received condition was about 10% stronger than the unglazed
showing the benefit of the compressive glaze.  Both materials were
strengthened by quenching in the glazed condition with the best
results observed by quenching from 1200°C.  The highest average
strength, 32,200 psi, is 35% higher than that of the unglazed
controls and 23% higher than the glazed controls.

## 2.4.7  Mullite

Gebauer and Hasselman (1971) measured the strength variations
of a mullite[/] ceramic subjected to thermal shock by quenching from

---

[*]ALSIMAG 475, 3M Company, Chattanooga, Tenn.

[+]HS-20, Lapp Insulator Division of Interpace, Leroy, New York.

[#]HS-20 (glazed).

[/]MV-20, McDanel Refractory Porcelain Co., Beaver Falls, Pa.

Table 2.22. Flexural Strength of Zircon Porcelain* Strengthened by Quenching (Cylindrical Rods 0.120 in diam)

| Treatment | No. Specimens | Average Flexural Strength+ psi | Strength Difference psi |
|---|---|---|---|
| Controls, as-received | 5 | 29,300 | --- |
| Controls, refired 1400°C | 5 | 29,300 | --- |
| Quenched from 1400°C into forced air | 5 | 32,200 | +2,900 |
| Glazed# at 1400°C and slowly cooled | 5 | 28,600 | - 700 |
| Glazed and quenched from 1400°C into forced air | 5 | 33,400 | +4,100 |

* ALSIMAG 475, 3M Company, Chattanooga, Tenn.

+ Four point loading on a two inch span.

# Regular glaze.

Table 2.23.  Flexural Strength of Electrical Porcelain Strengthened by Quenching (Cylindrical Rods ∿ 0.27 in diam)

| Treatment | No. Specimens | Average Flexural Strength[*] psi | Strength Difference psi |
|---|---|---|---|
| Lapp HS-20 | | | |
| Controls, as received | 4 | 23,900 | - - - |
| Glazed[+] and quenched from 1200°C into forced air | 3 | 30,700 | +6,800 |
| Glazed and quenched from 1300°C into forced air | 2 | 28,600 | +4,700 |
| Lapp HS (Glazed) | | | |
| Controls, as received (glazed) | 3 | 26,200 | - - - |
| Quenched from 1200°C into forced air | 3 | 32,200 | +6,000 |
| Quenched from 1300°C into forced air | 3 | 28,100 | +1,800 |

[*]Four-point loading on a two-inch span.

[+]G-24 frit.

various temperatures into silicone oil[*]. They observed a threshold
in the quenching temperature difference ($\Delta T$) at 750-800°C, above
which thermal shock damage was induced and the strength was degraded
as shown previously in Figure 2.4.   At higher $\Delta T$, the strength was
improved rather than degraded.   This improvement was attributed to
formation of compressive surface stresses due to elastic-plastic
phenomena at high temperatures during quenching.   The basic principle that it was necessary to quench from temperatures at which
creep rates were substantial in order to avoid thermal shock damage,
as illustrated in Section 2.1.4 for alumina, was also in use at
Ceramic Finishing Company.

Gebauer and Hasselman observed a strength increase of about
50% for mullite specimens quenched from about 1500°C.   Subsequently, Gebauer, Krohn, and Hasselman (1972) investigated thermal
stress fracture of similarly strengthened specimens.   The threshold
in the $\Delta T$, necessary to introduce thermal shock damage, was
increased to about 900°C as a result of the compressive surface
stresses.   Also, the remaining strength after thermal shock damage
is more than twice that of the specimens without the compressive
surface stresses.   This observation was explained on the basis that
the local stress at the surface (thermal stress minus residual
stress) is the same in both cases so that the strength degradation
should be approximately the same but because this degradation is
subtracted from a higher strength in the case of the strengthened
specimens the remaining strength is greater.   Actually, the
strength degradation of the strengthened specimens is approximately
39% greater than that of the original specimens.   It was concluded
that, to improve thermal stress resistance of ceramics, strengthening by surface compression is to be preferred to increasing the
inherent strength of the material.   The primary reason for this
is that increasing the inherent strength leads to increasingly
catastrophic failures.

---

[*]L-45, Union Carbide Corp., New York, N.Y.

## 2.5  Strengthening Non-Oxide Ceramics by
## Heating and Quenching

Two non-oxide ceramics, silicon carbide and silicon nitride, were strengthened by heating and quenching.  These materials are substantially more refractory than the oxide ceramics for which treatments were previously described so that higher treatment temperatures may be required, leading to some special problems.  Treatments for these materials are described in the following sections.

### 2.5.1  Strengthening Silicon Carbide by Heating

Strengthening of ceramics by thermal exposure at elevated temperatures for various periods of time has been observed in many investigations.  Possible strengthening mechanisms include flaw healing and relief of localized tensile stresses tending to wedge open surface flaws.  Despite this experience, strengthening of hot pressed silicon carbide[*] was observed unexpectedly during experiments in which the specimens were thermally exposed as controls for another experiment (Kirchner, 1974).  1/4 x 1/4 x 2-1/4 in specimens were heated in air at 1315°C for 50 hours.  There was no change in the appearance of the surfaces as a result of this treatment.  The impact resistances of the specimens were measured with the results shown in Table 2.24.  The average impact resistances of specimens tested at 25°C which were 4.4 and 4.8 in lbs can be compared with control values of 1.9 and 3.1 in lbs indicating an increase as a result of thermal exposure.  The average for specimens tested at 1315°C was 8.1 in lbs which can be compared with controls tested at 1325°C averaging 4.2 in lbs.

The similarity in the impact resistances of the thermally exposed specimens tested at 25°C and the controls tested at 1315°C

---

[*] NC-203 silicon carbide, Norton Company, Worcester, Mass.

Table 2.24.   Impact Resistance of Thermally Exposed* Silicon Carbide

| Specimen No. | Test Temp. °C | Impact Resistance[+] | | Comments |
|---|---|---|---|---|
| | | Joules | in lbs | |
| B-51-1C | 25 | 0.55 | 4.9 | Edge origin |
| B-51-2C | 25 | 0.44 | 3.9 | |
| B-51-3C | 25 | 0.48 | 4.3 | |
| Average | | 0.49 | 4.4 | |
| G-52-10 | 25 | 0.50 | 4.4 | Small internal mirror |
| G-52-11 | 25 | 0.67 | 5.9 | Corner origin |
| G-52-12 | 25 | 0.47 | 4.2 | Small internal mirror |
| Average | | 0.55 | 4.8 | |
| B-53-2C | 1315 | 1.06 | 9.4 | Corner origin |
| B-53-3C | 1315 | 0.77 | 6.8 | Corner origin |
| Average | | 0.92 | 8.1 | |

* Heated in air at 1315°C for 50 hours.
+ One foot pound hammer.

indicates that the controls tested at the elevated temperature may
benefit from the thermal exposure strengthening mechanism.

The observation of small internal fracture mirrors indicates
that, in at least two cases, the fractures originated at internal
flaws and that the fracture stresses were at least in the normal
range.  Extensive experiments (Kirchner, Gruver, and Sotter, 1976)
have shown that impact fractures in this as-machined silicon
carbide almost always originate at surface flaws.  Apparently,
thermal exposure treatment raises the stress at which surface
flaws act to cause failure thus permitting fractures to originate
internally.

Lange (1970) exposed hot pressed silicon carbide, containing
cracks formed by thermal shock, to air at 1400°C for various
periods of time, and measured the flexural strengths.  Even after
96 hours at 1400°C, the strength of the thermally exposed material
had increased only 5-10% over that of the thermally shocked material.
This increase was attributed to healing by oxidation of the surfaces
of the cracks.

The mechanism causing the improvement in impact resistance
described above may also involve healing of flaws by oxidation or
some other flaw healing mechanism but direct evidence is not available.   The results of Kirchner, Gruver, and Sotter (1975) have
shown that the impact energy absorbed during Charpy tests depends
primarily on the energy required to deflect the specimen to the
fracture stress which in turn varies as the square of the fracture
stress.  Therefore, the 84% increase in impact resistance indicated
by the room temperature data implies a 36% increase in strength.

## 2.5.2  Strengthening Silicon Carbide by Quenching

Because of the creep resistance of silicon carbide at elevated
temperatures one would expect that very high quenching temperatures
would be required to induce compressive surface stresses in hot
pressed silicon carbide.  Negative rod test deflections of specimens
quenched from temperatures ranging from 2000-2400°C into various

media indicate that compressive surface forces were induced
(Table 2.25) (Gruver, Platts, and Kirchner, 1974).  The results
are somewhat scattered so that no definite information was obtained
about the variation of these forces with quenching medium or
temperature.

Groups of cylindrical rods[*] were quenched from various temperatures into silicone oil (350 cSt) and the flexural strengths were
measured with the results shown in Figure 2.34.  Strengthening was
observed for groups of specimens quenched from 1950 and 2000°C but
the strengths of specimens quenched from higher temperatures were
degraded.  The highest average flexural strength was 98,700 psi,
a 25% increase compared with 79,000 psi for as-machined controls.
Other factors being equal, it is desirable to minimize exposure
of the specimens to high temperatures during treatment.

Larger specimens were degraded by thermal shock during quenching into silicone oils.  Therefore, less severe quenching media,
including fluidized beds, were used in a wide variety of experiments.  Cylindrical rods of hot pressed silicon carbide[+] were
quenched from 2000°C into a $ZrO_2$-air fluidized bed yielding an
average flexural strength of 126,000 psi (869 $MNm^{-2}$), which can
be compared with 99,900 psi for similar controls indicating a 26%
increase.

Rectangular bars (1/4 x 1/4 x 2-1/4 in) were quenched from
various temperatures into fluidized beds and the impact resistances
were measured.  Some groups of specimens had low values because
of thermal shock damage.  Excluding these, the average room
temperature impact resistance of quenched ACE-SiC was 3.8 in lbs
compared with control values averaging 2.3 in lbs.  At 1320°C
the average for quenched specimens was 3.9 in lbs compared with
2.9 in lbs for controls.  Average values for quenched NC-203 SiC
ranged from 2.0 to 4.5 in lbs which can be compared with average

---

[*] 0.1 in diam ACE SiC, Alfred Ceramic Enterprises, Alfred, N.Y.
[+] NC-203, Norton Company, Worcester, Mass.

Table 2.25.  Compressive Surface Forces as Indicated by Room Temperature Rod Tests (ACE SiC, 0.15 x 0.15 x 2.25 in)

| Quenching Medium | Quenching Temperature °C | Deflection mm |
|---|---|---|
| None, controls | None | None |
| Still air | 2000 | -0.06 |
|  | 2000 | -0.08 |
| 20 cSt silicone oil | 2130 | -0.15 |
|  | 2230 | -0.23 |
|  | 2230 | -0.15 |
|  | 2330 | -0.05 |
| 100 cSt silicone oil | 2130 | -0.19 |
|  | 2230 | -0.26 |
|  | 2330 | -0.19 |
|  | 2400 | -0.22 |
| $ZrO_2$-air fluidized bed | 2000 | -0.03* |
|  | 2000 | -0.09 |
|  | 2000 | -0.18 |
|  | 2200 | -0.21 |
|  | 2400 | -0.27 |
| SiC-air fluidized bed | 2000 | -0.19 |
|  | 2200 | -0.20+ |
|  | 2400 | -0.30+ |

*Specimen did not sink into quenching medium.
+Specimen was cracked.

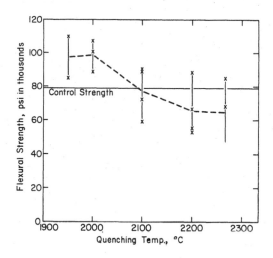

Figure 2.34.  Flexural strength of hot pressed silicon carbide
              quenched from various temperatures into silicone
              oil.  Reprinted with permission of Bull. Am. Ceram.
              Soc. 53 (7) (July, 1974), 524-527.

control values of 1.9 and 3.1 in lbs.  The weight of the above
evidence indicates that the quenching treatments improve the
impact resistance of hot pressed silicon carbide at 25 and 1320°C
when thermal shock damage is avoided.  The observed increases may
be less than those achieved simply by thermal exposure at much
more moderate temperatures.  Therefore, even though the compressive
surface stresses are present, little or no additional impact resistance was observed.

### 2.5.3  Strengthening Silicon Nitride by Heating

Cylindrical rods of hot pressed silicon nitride[*] were heated
to 1350°C and slowly cooled (Kirchner, Sotter, and Gruver, 1975)
with the results shown in Table 2.26.  The strength increase was

---

[*]NC-132, Norton Company, Worcester, Mass.

Table 2.26. Flexural Strength of Heated Silicon Nitride (Rods 3 mm diameter)

| Treatment | Flexural Strength $MNm^{-2}$ | Standard Deviation $MNm^{-2}$ | Average Flexural Strength $MNm^{-2}$ |
|---|---|---|---|
| Machined and polished controls | 821<br>821<br>778<br>757<br>653 | 61.7 | 766 |
| Heated to 1350°C and slowly cooled | 1142<br>1031<br>1023<br>978<br>434 | 248.2 *<br>(60.3) | 922 *<br>(1044) |

*Four highest values only.

20% or 36% depending on whether or not one unusually low value
was included in the average.

The observed increase in strength may occur as a result of
flaw healing or relief of localized residual tensile stresses at
flaws. Petrovic, Jacobson, Talty, and Vasudevan (1975) calculated
critical stress intensity factors ($K_{IC}$) at controlled surface flaws
in hot pressed silicon nitride after annealing for six hours at
various temperatures at 800°C and above. Low values of $K_{IC}$ in
unannealed specimens were explained on the basis that localized
residual tensile stresses tended to wedge open the flaws so that
failure to include these stresses in the calculation reduced the
calculated values of $K_{IC}$. After annealing at temperatures ranging
from 1200-1400°C, the localized stresses were relieved and the
$K_{IC}$ values returned to the normal levels. These authors discounted
the importance of flaw healing on the basis that 2:1 changes in
flaw size would be needed to account for the measured strength
variations and these changes in flaw size should have been easily
observed. Experience indicates that this argument is not conclusive.

Another factor may be rounding of the crack tip. Means to
measure changes in the radius of the crack tip are not available
so direct evidence is lacking.

It seems reasonable that if localized residual tensile
stresses are present at artificially induced flaws, they can also
be present at natural flaws introduced during machining. Therefore, at least part of the observed strength increase observed as
a result of heating can be attributed to relief of these stresses.

### 2.5.4 Strengthening Silicon Nitride by Quenching

Hot pressed silicon nitride specimens[*] were quenched from
various temperatures into various media and the flexural strengths
were measured by four point loading on a one inch span. The results

---

[*]NC-132 silicon nitride rods, 3 mm diam.

for specimens quenched from various temperatures into silicone
oil (100 cSt) are presented in Figure 2.35.  They are very
similar to those for silicon carbide in that specimens quenched
from lower temperatures (1350, 1400°C) were strengthened while
those quenched from higher temperatures were weakened.  Specimens
quenched from 1350 or 1400°C into silicone oils of various viscosities were also strengthened.  The highest values were obtained
by quenching from 1350°C into 50 cSt silicone oil.  The average of
these values was 1100 MNm$^{-2}$, about 30% higher than the control
values.

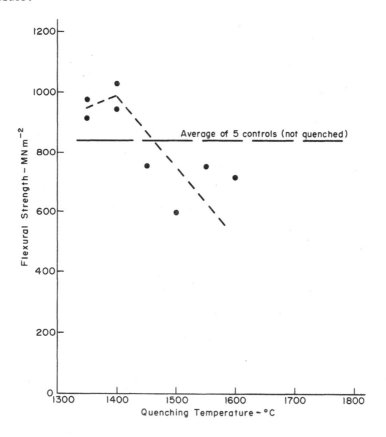

Figure 2.35.   Flexural strength of hot pressed silicone nitride
              quenched from various temperatures in silicone oil,
              100 cSt (four-point loading on a one-inch span).

To check the above results, additional specimens were quenched
from 1350°C into 50 cSt silicone oil.  In addition, specimens were
quenched in other quenching media and the possibility that the
strengthening was an artifact caused by protection from stress
corrosion by silicone oil was checked by measuring the strengths
of control specimens that were simply dipped into the silicone
oil before testing.  These results are given in Table 2.27.  The
strengths of controls that were dipped in silicone oil at room
temperature were not significantly different from those of the
undipped controls showing that the observed strengthening of the
quenched specimens was not simply due to protection from stress
corrosion.  All of the quenched groups were substantially stronger
than the controls.  The highest individual strength value was
1226 MNm$^{-2}$ (178,000 psi) which was observed for a specimen quenched
into water plus emulsifier.

Tensile tests were performed using necked down specimens,
1-2 mm in diam quenched from 1350°C into silicone oil (50 cSt),
with the results shown in Table 2.28.  The average tensile strength
of the quenched specimens was 829 MNm$^{-2}$ compared with 700 MNm$^{-2}$
for the controls, an 18% increase.  In both controls and treated
specimens the fractures originated internally.  Therefore, the
observed strengthening cannot be attributed to healing of flaws
introduced during machining.

The residual stresses induced by quenching from 1350°C into
silicone oil (50 cSt) were estimated by the fracture mirror method
described in Section 1.3.3.  The mirror radii of the quenched
flexural specimens in which the fractures originated at surface
flaws were measured and compared with similar measurements of the
control specimens (Figure 2.36).  Because the size of the mirror
depends on the local stress at fracture, the vertical distance
between the curves of Figure 2.36 can be used to estimate the
residual compressive surface stress which, for a mirror radius of
$10^{-4}$ m is $\simeq$ 120 MNm$^{-2}$.  This result suggests that 1/3 to 1/2 of the
strength increase of the quenched flexural specimens arises from
surface residual compressive stresses.

Table 2.27.  Flexural Strength of Silicon Nitride Rods with Various
             Treatments

| Treatment | Flexural Strength $MNm^{-2}$ | Average Flexural Strength $MNm^{-2}$ | Standard Deviation $MNm^{-2}$ |
|---|---|---|---|
| Controls, machined and polished | 821 | | |
| | 757 | | |
| | 821 | | |
| | 778 | | |
| | 653 | 766 | 61.7 |
| Dipped in 50 cSt silicone oil, washed in cold water | 762 | | |
| | 900 | | |
| | 804 | | |
| | 584 | | |
| | 852 | 780 | 109.7 |
| Dipped in 50 cSt silicone oil, not washed | 656 | | |
| | 756 | | |
| | 745 | | |
| | 836 | | |
| | 788 | 756 | 61.7 |
| Quenched from 1350°C into water | 1029 | | |
| | 687 | | |
| | 1071 | | |
| | 891 | | |
| | 1101 | 956 | 151.2 |
| Quenched from 1350°C into water plus emulsifier | 1148 | | |
| | 853 | | |
| | 1163 | | |
| | 716 | | |
| | 1226 | 1021 | 200.8 |
| Quenched from 1350°C into forced air | 965 | | |
| | 890 | | |
| | 1050 | | |
| | 968 | | |
| | 1022 | 979 | 55.0 |
| Quenched from 1350°C into 50 cSt silicone oil | 1060 | | |
| | 1048 | | |
| | 975 | | |
| | 1087 | | |
| | 1103 | 1055 | 44.3 |

Table 2.28.   Tensile Strengths of Quenched Silicon Nitride

| Treatment | Tensile Strength MNm$^{-2}$ | Average Tensile Strength MNm$^{-2}$ | Standard Deviation MNm$^{-2}$ | Flaw Characterization | | |
|---|---|---|---|---|---|---|
| | | | | Type | Size µm | Distance from Surface µm |
| Controls | 896 | | | Porous aggregate | 6 | 532 |
| | 716 | | | Porous region | 7 | 68 |
| | 708 | | | Porous region | 5 x 10 | 280 |
| | 651 | | | Porous region | 5 x 10 | 370 |
| | 529 | 700 | 118.6 | Porous region | 10 x 20 | 15 |
| Quenched from 1350°C into silicone oil 50 cSt | 925 | | | Porous aggregate | 6 | 470 |
| | 911 | | | Porous region | 2 | 346 |
| | 755 | | | Porous region | 5 x 30 | 98 |
| | 726 | 829 | 91.8 | Porous region | 6 | 143 |

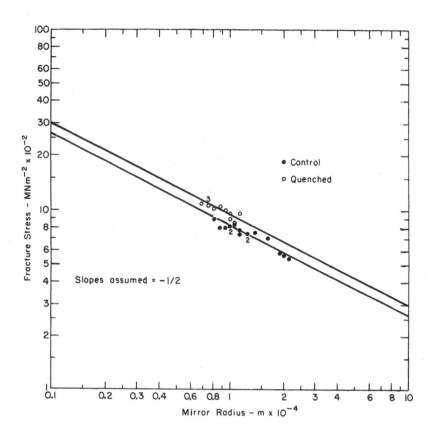

Figure 2.36. Fracture stress vs. mirror radius for hot pressed
           silicon nitride controls and specimens quenched from
           1350°C. Reprinted with permission of J. Am. Ceram.
           Soc. 58 (7-8) (1975), 353.

Comparing the results obtained by quenching with those
reported in the last section for specimens that were heated but
not quenched indicates that most of the strengthening reported for
the quenched specimens can be attributed to changes occurring due
to heating.  The nature of these changes is uncertain because all
of the fracture origins in the tensile specimens, both controls and
quenched, were at internal flaws.  The residual tensile stress
mechanism of wedging open the flaws seems improbable for internal

flaws which must have been subjected to high temperatures during
hot pressing so that one would expect any localized stresses to be
relieved. There is no obvious mechanism by which these flaws
could be altered subsequent to hot pressing because they are protected by the surrounding material.

The residual compressive surface stresses in specimens quenched
from 1350°C into silicone oil (50 cSt) are about 120 $MNm^{-2}$. Previous
experience with a wide range of other materials indicates that these
stresses are responsible for strengthening. Therefore, until other
evidence is obtained it seems reasonable to attribute 1/3 to 1/2
of the observed strengthening in the quenched specimens to strengthening by compressive surface stresses and the balance to strengthening by an unknown mechanism induced by heating alone.

## 2.6 Status of Research on Thermal Treatments

There are several questions frequently raised about strengthening of ceramics by thermal treatments, such as:

1. Isn't the observed strengthening simply a result of flaw
   healing or relief of localized stresses by annealing?
2. Is it really possible to induce compressive surface stresses
   in ceramics by quenching?
3. Even if it is possible to use compressive surface stresses to
   raise the nominal stress at which surface flaws, act to cause
   failure, won't the internal flaws become fracture origins
   under the combined influence of the residual tensile stresses
   and stresses due to the applied loads, preventing useful
   increases in load carrying ability?
4. Isn't a siliceous intergranular phase necessary in order to
   induce compressive surface stresses by quenching?
5. Won't the application of ceramics strengthened by quenching,
   at high temperatures, be prevented by relief of the compressive stresses?

Some answers to these questions are available and will be discussed
in the following paragraphs.

In many cases refired controls were included in the quenching
experiments. In some cases these specimens were strengthened,

presumably by crack healing, crack tip rounding or by relief of
localized residual tensile stresses at surface damage.  In the case
of 96% alumina, for example, this strengthening effect was observed
many times but the magnitude of the effect was very small in comparison with the increase in strength obtained by quenching.  Also,
the distribution curves (Section 2.1.5) are inconsistent with those
obtained by quenching.  One exception is sapphire for which large
increases in strength were obtained by refiring and slow cooling.
Even in this case quenching may still be an advantage because any
stresses that may be induced tend to protect the surface from
damage.

The presence of the compressive surface stresses has been
demonstrated by ring tests, rod tests, x-ray diffraction measurements of elastic strain, and by fracture mirror measurements.  In
addition, Buessem and Gruver (1972) calculated the stress profiles
to be expected from creep, heat transfer and thermal expansion data.
There is a strong correlation between the strength improvements and
the relative magnitudes of the compressive surface forces as
indicated by the rod tests.  For these reasons, there is no question that compressive surface stresses can be induced in ceramics
by quenching and, in many cases they are responsible for most of
the observed strengthening.

There is strong evidence presented in the previous sections
that despite the shift from surface to internal fracture origins
when compressive surface stresses are induced by thermal treatments, substantial increases in load carrying ability are achieved.
The best evidence of this is the improved tensile strength because
interpretation of these results does not require precise knowledge
of the residual stress profile and its effect on the maximum
stresses as is the case for flexural strength measurements.
Whether or not the shift from surface to internal fracture origins
will prevent substantial improvements in load carrying capacity
depends primarily on the relative severity of the flaws in the
two types of locations.  As the quality of ceramic bodies improves,
internal flaws can be expected to become even less important whereas

surface flaws can be expected to continue to be critical because
of surface damage in use.  Therefore, as time passes, the usefulness of compressive surface stresses should increase.

Because of analogies with thermal tempering of glass and
because early work on "thermal conditioning" of ceramics was done
using bodies with siliceous intergranular phases, it was frequently
assumed that the presence of a siliceous (or viscous) intergranular
phase was necessary to induce compressive surface stresses in
ceramics by quenching.  More recently, several materials with
little or no viscous intergranular material, including hot pressed
alumina and silicon carbide have been strengthened by quenching.
In addition, it is now well known that extensive plastic deformation can occur in single phase ceramics at high temperatures.
Therefore, a viscous intergranular phase is not necessary to make
it possible to strengthen ceramics by quenching.

The question of application of ceramics, strengthened by
quenching, at high temperatures is for the most part unanswered
at present.  Gebauer, Krohn, and Hasselman (1972) have shown that
compressive surface stresses induced by quenching are effective
in raising the temperature difference necessary to cause failure
of mullite during subsequent more severe quenching.  Kirchner and
Gruver (1974) held 96% alumina specimens strengthened by quenching
at 850 and 900°C for four hours and observed strengths similar to
specimens that were tested immediately at those temperatures.
These strengths were substantially greater than those of unquenched
specimens tested under the same conditions.  Both investigations
indicate potential usefulness of quenched ceramics at elevated
temperatures.

Arguments for the usefulness of quenched ceramics at elevated
temperatures depend on the difference between the quenching
temperature and the prospective use temperature.  For example,
96% alumina might be quenched from 1500°C and considered for use
at 800°C.  Hot pressed silicon carbide might be quenched from
2000°C and considered for use at 1400°C.  Because of the exponen-
tial dependence of creep rates on temperature, the time required

to relieve stresses at the use temperature may be many times the
time required to induce them at the quenching temperature.

If one assumes that the creep rate $\dot{\epsilon} \propto e^{\frac{-1}{T}}$ in which T is
the absolute temperature and that the mechanism of creep remains
the same in 96% alumina from 1500°C to 900°C, one can extrapolate
the high temperature creep data of Buessem and Gruver (1972) to
obtain an estimate of the creep rate at 900°C which is about
$\frac{1}{5700}$ that at 1500°C. Although analysis of stress relaxation is
a complex problem it is evident that stresses induced in seconds
or minutes at 1500°C will take at least several hours to be
relieved at 900°C.

### 2.6.1  Summary of Successes and Failures

Not all treated materials have been strengthened by quenching.
The successes and failures are summarized in Table 2.29. The
materials listed as not yet strengthened by quenching, should not
be considered as "lost causes" because, in many cases, the experiments were very limited. Except for barium titanate there seems
to be no fundamental difference between the materials that were
strengthened and those that were not. Barium titanate is a
piezoelectric material and the particular body tested was coarse
grained and rather weak to begin with. Failure to strengthen this
material may have been caused by lack of resistance to thermal
shock as a result of the coarse grained structure.

### 2.6.2  Effect of Various Shapes and Sizes on
### Strengthening Results

Development of techniques for quenching specimens of various
shapes and sizes is a substantial problem. The simplest case is
quenching of spheres for which the temperatures and stresses are
radially symmetrical in three dimensions and there are no corners
or edges at which thermal shock cracks can initiate. Cylindrical
rods are slightly more complex because of the discontinuities at

Table 2.29.  Summary of Successes and Failures in Quenching Experiments[*]

| Materials Strengthened by Quenching | Materials Not Yet Strengthened by Quenching |
| --- | --- |
| 96% alumina ($Al_2O_3$) | Magnesium oxide (MgO) |
| H.P. alumina ($Al_2O_3$) | Barium titanate ($BaTiO_3$) |
| Sapphire single crystals ($Al_2O_3$) | Cordierite ($Mg_2Al_4Si_5O_{18}$) |
| Titania ($TiO_2$) | Forsterite ($Mg_2SiO_4$)[+] |
| Spinel ($MgAl_2O_4$) | |
| Steatite ($MgSiO_3$) | |
| Zircon porcelain ($ZrSiO_4$) | |
| Electrical porcelain | |
| Mullite ($Al_6Si_2O_{11}$) | |
| Silicon carbide (SiC) | |
| Silicon nitride ($Si_3N_4$) | |
| Forsterite ($Mg_2SiO_4$)[+] | |

[*] Of course there are many ceramics that, as yet, have not been investigated.

[+] Forsterite is listed in both groups because it has strengthened only relative to refired specimens.

the ends.  Occasionally, thermal shock cracks originate at the
ends and propagate axially the entire length of the specimen.
In alumina these cracks must originate early in the cooling because
they appear to heal and cause little or no loss of strength.  Formation of these cracks can usually be prevented by rounding the
edges at the ends or by using coatings or other means to reduce
the cooling rate at the ends.  Hollow cylinders are a still more
severe problem because of the two sets of edges at each end and
the inner surface which usually cools at a lower rate than the outer
surface.  Experimental evidence describing the stresses in the inner
surface is lacking but they are probably much lower than those in
the outer surface and they may even be tensile stresses.  When
tested in flexure the applied stress at the inner surface may be
low enough so that fracture does not originate there but in
tensile tests this surface is likely to be vulnerable.

Rectangular bars can be expected to cause difficulties for
two reasons  (1) the edges cool rapidly leading to high tensile
stresses early in the cooling process and  (2) the edges are
likely to contain flaws at which thermal shock cracks may originate.
Usually, it is necessary to radius the edges to reduce the cooling
rate and to reduce the severity of the edge flaws.

Quenching of plates presents problems similar to those of
rectangular bars.  Very thin plates such as substrates for micro-
circuits have been quenched.  Thicker plates including armor tiles,
6 x 6 x 3/8 in, have also been quenched without evident damage.

A recent bibliography on the thermal stress fracture of
ceramics, glasses and refractories can serve as a source of information to use in solving thermal shock problems (Specian and Hasselman, 1975).

Hasselman (1970) discussed the effect of the radius of
cylindrical rods on the thermal stress, elastic energy at fracture and remaining strength after thermal shock damage in alumina
ceramics.  The maximum thermal stress ($\sigma_{max}$) can be estimated,
based on Glenny and Royston (1958) and assuming no plastic deformation, using

$$\sigma_{max} = B \frac{\alpha E}{(1 - \nu)} \Delta T \tag{2.3}$$

in which B (Biot's modulus) is a function of bh/k where b is
the rod radius, h is the heat transfer coefficient and k is the
thermal conductivity, $\alpha$ is the coefficient of thermal expansion,
E is Young's modulus, $\nu$ is Poisson's ratio, and $\Delta T$ is the quenching
temperature difference. Increases in b and h increase $\sigma_{max}$ whereas
increases in k decrease it. The elastic energy (W) at fracture,
per unit length of specimen, is

$$W = 0.57 \frac{\sigma_f^2 b^2}{E} \tag{2.4}$$

in which $\sigma_f$ is the fracture stress. Assuming that the elastic
energy is used to form fracture surfaces, the Griffith equation
can be used to calculate the strength after thermal shock damage
$(\sigma_a)$ which is

$$\sigma_a = \left( \frac{8\gamma_t \gamma_s^2 E^3 N_s}{\sigma_f^2 b} \right)^{1/4} \tag{2.5}$$

in which $\gamma_t$ is the fracture energy appropriate for the thermal
shock environment, $\gamma_s$ is the fracture energy appropriate for the
strength test and $N_s$ is the number of cracks per unit area.
Hasselman found that $\sigma_a \propto (\frac{1}{b})^{1/4}$ as predicted by this equation.

Only a small amount of data is available to show how the
effectiveness of quenching varies with rod diameter. One expects
that at small rod diameters the interior might cool before sufficient plastic deformation occurs so that larger diameter rods
would show increased residual stress and strength. On the other
hand, at very large diameters, it may be necessary to reduce the
cooling rate to avoid thermal shock damage, so that quenching may
be less effective. Figure 2.37 gives results for as received and
quenched specimens of five different diameters. A principal problem
in experiments of this type is to fabricate specimens of comparable

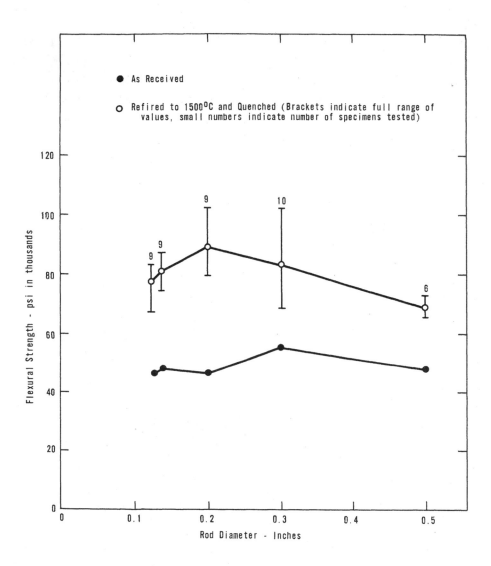

Figure 2.37.  Flexural strength vs. rod diameter for 96% alumina
rods quenched from 1500°C into silicone oil (12,500
cSt).  Reprinted with permission of J. Appl. Phys.
42 (10) (1971) 3685-3692.

strength over the entire range of sizes.  In this particular case
the strengths of the 96% $Al_2O_3$ rods with diameters ranging from
0.125 to 0.50 in are considered to vary within reasonable limits
in the as received condition.  In the quenched condition, the
average strength increased with increasing rod diameter to 0.2 in
and then decreased as might have been expected.

### 2.6.3  Summary of Benefits of Quenching

Experiments have shown that flexural strength, tensile strength,
thermal shock resistance, impact resistance, resistance to delayed
fracture and resistance to penetration of surface damage can be
improved by quenching.  The only alternative fundamental methods
of strengthening of brittle materials are to reduce the flaw size
and increase the fracture toughness.  Assuming that ceramics will
be subjected to surface damage in most practical applications means
that reduction of flaw size can yield only transient benefits.  On
the other hand, compressive surface stresses which in effect raise
the load that can be carried at each level of surface stress can
assure reliability despite surface damage because, when surface
damage does occur, it can be expected to be less severe.

The roles of subcritical crack growth and proof testing in
assuring the reliability of ceramics have been emphasized recently
(Wiederhorn, 1974 a,b) and are certain to be decisive in determining
the scope of future structural applications of ceramics.  Compressive surface stresses substantially reduce the stress intensity
factors and crack growth rates at surface flaws resulting in
extension of the times to failure at a particular load by many
orders of magnitude.  This advantage is illustrated by the results
of Kirchner and Walker (1971) in which the delayed fracture tests
of quenched 96% $Al_2O_3$ can be considered the equivalent of proof
tests for components subjected to bending stresses at room temperature.

## 2.6.4  Recommendations for Research

A basic requirement for progress in strengthening of ceramics by compressive surface stresses is improvement of the bodies to which the treatments are applied.  If the severity of internal flaws in polycrystalline ceramics can be reduced to approach those in glass, the effectiveness of the compressive surface stresses should be much improved.  In addition to this basic improvement there are several areas in which available information is insufficient for evaluation of quenching as a method of strengthening ceramics for practical applications.  The following are some of these areas:

1.  Compare the tensile and flexural strengths of quenched speci-
    mens to determine the relative importance of compressive
    surface stresses in increasing the stress at which surface
    flaws act to cause failure and in reduction of tensile
    stresses at internal flaws in flexural specimens as a result
    of shifting of the stress profile.
2.  Determine the rates at which the compressive surface stresses
    are relieved at various elevated temperatures and the effect
    of the relief of these stresses on the strength.
3.  Improve the methods of calculating the stress profiles in
    quenched specimens using creep, heat transfer and thermal
    expansion data.
4.  Prepare an analysis of this strengthening technique using a
    fracture mechanics approach.
5.  Demonstrate the application of new techniques of proof testing
    and evaluation of subcritical crack growth to determination
    of the reliability of quenched ceramics for structural appli-
    cations.

In addition to the research described above, process development should continue to improve the properties of materials previously strengthened and to adapt this technique to strengthening of new materials.

# CHAPTER 3

# COATINGS AND CHEMICAL TREATMENTS

## 3.1  GLAZES

As mentioned previously, low expansion glazes have been used to strengthen ceramics since ancient times.  Recently, glazes have been used to improve the impact resistance of glass-ceramic cylinders intended for use in underwater vehicles for deep submergence (Conway, 1971).  In the following sections, techniques for strengthening modern ceramics by use of low expansion glazes and by ion exchange of glazes are presented.

### 3.1.1  Low Expansion Glazes

The compositions and thermal expansion coefficients of the "regular," L-2 and S-1 glazes were given previously in Table 2.3. 96% alumina rods, 0.125 in diam, were glazed with these compositions and slowly cooled (Kirchner, Gruver, and Walker, 1968).  The thermal expansion coefficient of the 96% alumina body is $65 \times 10^{-7} \degree C^{-1}$ (25-300°C) and the coefficients of the glazes fall above and below this value.  Because of changes in composition due to vaporization and reaction with the body, the expansion coefficients of high expansion glazes are expected to tend toward that of the body.  This tendency will be greater for thin glazes than for thick glazes.

The rod test data, flexural strengths and other relevant
information are assembled and compared in Table 3.1.  The rod
test results show that the 96% alumina rods are damaged during
slotting yielding an increase in rod diameter of 0.002 in.  After
glazing with the various glazes, the rod test results are consistent
with expectations based on the relative thermal expansion coefficients.  The "regular" glaze with the lowest expansion coefficient
yields the greatest negative rod test deflection and the greatest
improvement in strength relative to both as received and refired
specimens.  The L-1 glaze has an expansion coefficient greater
than the body but yields a negative rod test deflection.  Thick
layers on the alumina craze as might be expected based on the
relationship of the thermal expansion coefficients.  However, in
thin layers a larger proportion of the sodium is lost by evaporation and by reaction with the body, tending to reduce the expansion
coefficient.  Apparently, the resulting thermal expansion- coefficient is slightly less than that of the body yielding a substantial
improvement in strength.  The S-1 glaze has an expansion coefficient
much greater than the body yielding a positive rod test deflection,
tensile stresses in the glaze and reduced strength.

The S-4 glaze has a thermal expansion coefficient of
$73.5 \times 10^{-7}{}^{\circ}C^{-1}$, a value higher than that of the 96% alumina body.
This glaze was matured at various temperatures in the range
1250-1500°C.  The flexural strengths increased from 43,800 psi
for specimens fired at 1250°C with no hold period to 72,100 psi
for specimens fired at 1500°C with a one hour hold period.  Thus,
there is an increase in strength with increasing glaze maturing
temperature.

The results presented in this section show that low expansion
glazes are effective in inducing compressive surface stresses and
in improving the strength.

Table 3.1. Strengthening by Low Expansion Glazes on 96% Alumina* (Thermal expansion coefficient 65 x 10⁻⁷°C, 25-300°C)

| Glaze | Treatment | Rod Test Average Change in Rod Diam in | Thermal Exp. Coef. 25-300°C °C⁻¹x10⁷ | Flexural Strength Data[+] | | |
|---|---|---|---|---|---|---|
| | | | | No. Specimens | Average Strength psi | Strength Increase psi |
| None | None | --- | --- | 19 | 49,700 | --- |
| None | Refired and slowly cooled 1500°C, one hour | +0.002 | --- | 5 | 54,600 | +4,900 |
| Regular | Glazed and slowly cooled 1500°C, one hour | -0.001 | 53 | 5 | 70,800 | +21,100 |
| L-1 | Glazed and slowly cooled 1500°C, one hour | -0.0005 | 75 | 5 | 70,000 | +20,300 |
| S-1 | Glazed and slowly cooled 1500°C, one hour | +0.004 | 103 | 5 | 42,300 | - 7,400 |

*0.125 in diam rods.

[+]Four-point loading on a two-inch span.

## 3.1.2  Strengthening Glazed Alumina by Ion Exchange

Processes for stuffing glass surfaces to induce compressive
surface stresses are well known (Kistler, 1962; Stookey, 1965).
Usually, larger ions from a fused salt bath or other source are
exchanged for smaller ions in the glass surface at temperatures
below the strain point.  The glass network does not relax, but
instead expands to accommodate the larger ions.  Since the surface
is restrained by the underlying glass this leads to compressive
surface stresses.

The glazes used to demonstrate this process are described in
Table 3.2 (Platts, Kirchner, Gruver, and Walker, 1970).  The HL
glaze is a low expansion lithium aluminum silicate composition
and its thermal expansion coefficient was not measured.  In the
sodium aluminum silicate system, compositions with high sodium
contents and the best ion exchange properties have thermal expansion coefficients that are too high to fit alumina bodies properly.
Therefore, three different compositions were tested.  The lithium
and sodium ions were exchanged by immersing the glazed specimens
in molten salts of sodium and potassium respectively and the
strengths were measured with the results shown in Table 3.3.

### HL glaze

96% alumina rods were glazed with the HL glaze and ion
exchanged in $NaNO_3$.  The flexural strength was improved by ion
exchange and increased with increasing treatment time.

### S1 glaze

Greater improvements in strength were observed for 96% alumina
rod glazed with the S1 glaze and ion exchanged in $KNO_3$.  The strength
increased to a peak of 75,900 psi after 60 minutes treatment and
then declined.  Rod tests show that the glaze is in tension before
ion exchange and still in slight tension after ion exchange.  It

Table 3.2. Ion Exchange Glazes for Alumina Ceramics

| Glaze Designation | Composition wt % | | | | | | Thermal Expansion Coefficient (25-300°C) °C x 10$^7$ |
|---|---|---|---|---|---|---|---|
| | Li$_2$O | Na$_2$O | B$_2$O$_3$ | Al$_2$O$_3$ | SiO$_2$ | | |
| HL | 11.8 | --- | --- | 40.5 | 47.7 | | --- |
| S1 | --- | 19.5 | --- | 32.5 | 48.0 | | 103.0 |
| S3 | --- | 15.1 | 2.0 | 24.9 | 58.0 | | 70.8 |
| S4 | --- | 15.9 | --- | 26.2 | 57.9 | | 73.5 |

Table 3.3.   Strengthening of Glazed 96% Alumina* by Ion Exchange

| Glaze | Ion Exchange Condition | | | Flexural Strength Data[+] | | | |
|---|---|---|---|---|---|---|---|
| | Time °C | Time min | Salt | No. Specimens | Average Flexural Strength psi | Strength Increase psi | Rod Test Results |
| None | --- | --- | --- | --- | 44,800 | --- | --- |
| As received controls | | | | | | | |
| HL glaze 1500°C, 60 min | None | | | 3 | 44,400 | - 400 | --- |
| Same | 400 | 15 | $NaNO_3$ | 3 | 48,700 | + 3,900 | --- |
| Same | 400 | 30 | same | 3 | 52,400 | + 8,000 | --- |
| S1 glaze 1500°C, 60 min | None | | | 5 | 42,300 | - 2,500 | Tension |
| Same | 400 | 15 | $KNO_3$ | 4 | 59,600 | +14,800 | --- |
| Same | 400 | 30 | same | 4 | 64,700 | +19,900 | --- |
| Same | 400 | 60 | same | 4 | 75,900 | +31,100 | Slight tension |
| Same | 400 | 120 | same | 4 | 71,500 | +26,700 | --- |
| S3 glaze 1500°C, 60 min | None | | | 5 | 75,200 | +30,400 | --- |
| Same | 375 | 60 | 75% $KNO_3$ 25% $K_2SO_4$ | | 71,000 | +26,200 | --- |
| Same | 400 | 60 | same | 2 | 71,900 | +27,100 | --- |
| Same | 450 | 60 | same | 2 | 83,400 | +38,600 | --- |
| Same | 500 | 60 | same | 2 | 79,500 | +34,700 | --- |
| Same | 550 | 50 | same | 2 | 72,500 | +27,700 | --- |
| S4 glaze 1250°C, 60 min | None | | | 4 | 42,600 | - 2,200 | Slight tension |
| Same | 430 | 9 | 97% $KNO_3$ 3% $K_2SO_4$ | 4 | 54,500 | +9,700 | No stress |
| Same | 430 | 36 | same | 4 | 60,700 | +15,900 | No stress |
| Same | 430 | 64 | same | 4 | 63,000 | +18,200 | Slight compression |

*ALSIMAG 614, 3M Co., Chattanooga, Tenn.  0.125 in diam rods.
+Four point loading on a two inch span.

seems probable that, even though the glaze as a whole may be in
slight tension after ion exchange, the surface of the glaze which
contains the highest concentration of the larger potassium ions is
in compression.

## S3 glaze

96% alumina rods were glazed with the S3 glaze (1500°C, one
hour) and ion exchanged in 75% $KNO_3$ + 25% $K_2SO_4$ at 500°C for various
time intervals. The strength was increased substantially by glazing
alone. Further improvements were observed as a result of ion
exchange for various periods of time with the maximum strength
just over 74,000 psi after 180 min treatment. The strength after
60 min, 73,000 psi, was only slightly less than that after the
longer treatments. Based on these results, a treatment time of
60 min was chosen and temperature of the ion exchange process was
varied from 375 to 550°C. The flexural strength reached a maximum
for treatments at 450°C. At higher treatment temperatures the
network tends to relax decreasing the stress. Tensile specimens
were treated at 500°C for 60 minutes with the results shown in
Table 3.4. The tensile strength of the treated specimens was
59,300 psi which represents an improvement over as received
controls and glazed controls.

## S4 glaze

The effect of glaze maturing temperature was investigated
using the S4 glaze. The strength of the glazed specimens increases
with increasing time and temperature as shown in Section 3.1.1.
Moderate increases in strength were observed when specimens glazed
at 1250 and 1300°C were ion exchanged in $KNO_3$ at 375°C for short
periods of time. Another group of specimens was glazed at 1250°C
and ion exchanged in 97% $KNO_3$-3% $K_2SO_4$ for various periods of time
at 430°C with the results given in Table 3.3. These results show
increasing strength with increasing ion exchange time under these

Table 3.4.   Tensile Strength of 96% Alumina Glazed with S3 and Ion Exchanged (Rods 0.198 diam necked down by grinding)

| Treatment | No. Specimens | Average Tensile Strength psi |
|---|---|---|
| Controls, as received | 11 | 39,800 |
| Glazed with S3 at 1500°C for 60 min | 3 | 53,800 |
| Glazed with S3 at 1500°C for 60 min and ion exchanged in 75% $KNO_3$ -25% $K_2SO_4$ for 60 min at 500°C | 4 | 59,300 |

conditions. Rod tests of similarly treated specimens show increasing compressive surface stress with increasing treatment time, as expected. Specimens glazed at 1500°C were not strengthened by ion exchange, perhaps because the surface of the glaze was depleted in sodium as a result of evaporation or reaction with the interface. Electron microprobe analyses were used to investigate the distribution of alkali in specimens glazed at 1250 and 1500°C with and without the ion exchange treatment (Figure 3.1) showing substantial exchange of potassium for sodium in the specimens glazed at 1250°C and depletion of the sodium at 1500°C so that little ion exchange occurred confirming the above expectation.

The experimental results described in this section show that stuffing processes usually used for strengthening glass can also be used for strengthening ceramics. This was done by the ion exchange glazing technique.

## 3.2 LOW EXPANSION SOLID SOLUTION SURFACE LAYERS

Low expansion solid solution surface layers can be formed by diffusing an appropriate element into a ceramic body at a high temperature. Compressive surface stresses are induced during subsequent cooling because the interior contracts more than the surface, thus placing the surface in compression.

This chemical approach to formation of compressive surface stresses depends on the availability of thermal expansion data for solid solutions. Much useful information was published by Kirchner, Scheetz, Brown, and Smith (1962), Merz, Brown, and Kirchner (1962), Kirchner and Gruver (1965), and Kirchner (1969), establishing the basis for development of these strengthening processes. Although the principal objective of some of these investigations was to reduce the thermal expansion anisotropy of oxides by solid solution modifications, an important by-product was information about a number of systems in which the solid solutions had lower volume expansion coefficients than at least one of the pure phases. Examples of such data are given in Figure 3.2.

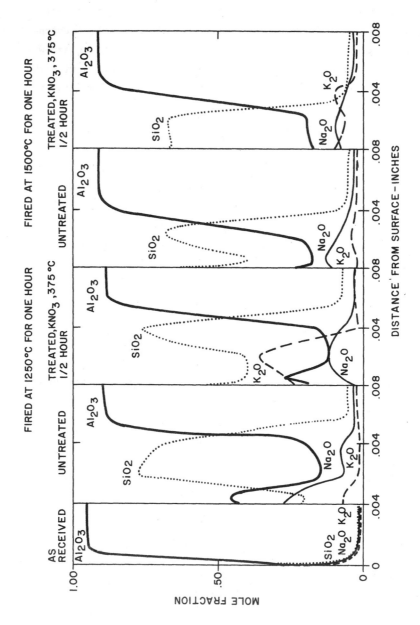

Figure 3.1.   Composition profiles for the S4 glaze on 96% alumina fired at 1250°C and 1500°C and ion exchanged.

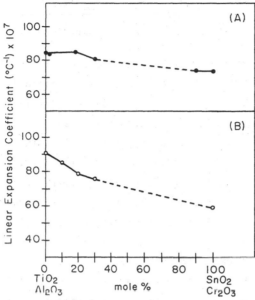

Figure 3.2.    Average linear thermal expansion coefficients of oxide
solid solutions, room temperature to 1000°C.    (A)
$Al_2O_3$-$Cr_2O_3$.  (B) $TiO_2$-$SnO_2$.  Reprinted with permission
of J. Am. Ceram. Soc. 49 (6) (1966) 330-333.

In most cases the solid solution surface layers were formed
by packing the ceramic body in a powder containing the desired
element and reheating for a substantial period of time.  Experience
has shown that it is necessary to form uniform surface layers.  This
can be done by using volatile reactants so that the surface is
exposed uniformly to the reactant.  If the process depends on
point contact between the powders and the surface, non-uniform
reaction occurs and the specimens are usually weaker.  The diffusion of the elements into ceramics is very slow and the temperature and treatment time are limited by factors such as grain
growth so that the low expansion solid solution surface layers
tend to be rather thin.  Thicker layers can be formed in some cases
by impregnating unfired, semi-fired, or leached bodies followed
by firing to obtain the desired final density.

The stresses expected in the surfaces of ceramic bodies with
low expansion solid solution surface layers can be calculated
using the following equation:

$$\sigma = \frac{E\Delta T}{1-\nu} \, (\alpha_{avg} - \alpha_{surf}) \qquad\qquad (3.1)$$

in which $\sigma$ is the compressive stress at the outer surface, E is
Young's modulus, $\Delta T$ is the temperature range of cooling after
firing, $\nu$ is Poisson's ratio, $\alpha_{surf}$ is the thermal expansion
coefficient of the surface material for $\Delta T$ and $\alpha_{avg}$ is the volume
averaged thermal expansion coefficient for $\Delta T$.  The calculations
were based on the following assumptions:  (1) A flat, infinite
plate coated on both sides;  (2) Surface layers 1/10 the thickness
of the plate;  (3) A linear composition gradient from the surface
of the plate to the interface between the solid solution layers
and the main body;  (4) A linear variation of thermal expansion
coefficient with composition;  (5) No relief of stresses during
cooling from the firing temperature;  (6) The thermal expansion
coefficients measured over a particular temperature range were
used at higher temperatures (this assumption tends to reduce the
difference $\alpha_{avg} - \alpha_{surf}$ in most cases, reducing the estimate of
$\sigma$); and  (7) The elastic properties of the solid solution are the
same as the elastic properties of a pure component.

This method of calculation takes into account the variation
of thermal expansion coefficient with position, the biaxial nature
of the stresses at the surface and the  Poisson's contraction.
The results of some calculations are given in Figure 3.3, indicating that compressive stresses sufficient for reasonable
strengthening can be achieved by this means.

The surface layers should be thick enough so that flaws
expected from surface damage do not penetrate the compressive
layer.  Nordberg, Mochel, Garfinkel, and Olcott (1964) reported
that a thickness of at least 80 μm was necessary to retain
flexural strength values above 50,000 psi in artificially abraded
glasses.  On the other hand, the compressive surface layers should
not be so thick that they result in substantial tensile stresses
in the core.

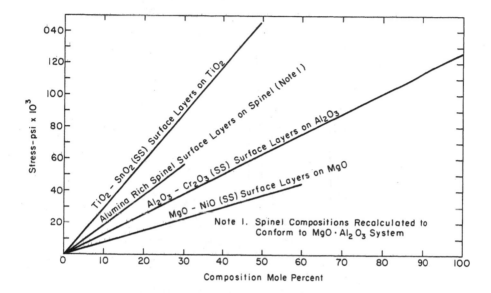

Figure 3.3.   Estimated compressive surface stress vs. composition
of surface layer for various material combinations.
Reprinted with permission of J. Am. Ceram. Soc.
49 (6) (1966), 330-333.

### 3.2.1 Polycrystalline Alumina

$Al_2O_3$-$Cr_2O_3$ solid solution surface layers were formed on
polycrystalline alumina ceramics by several techniques (Kirchner
and Gruver, 1966).  Rectangular bars, 1/4 x 9/16 x 3 in, were cut
from 96% alumina tiles[*].  The bars were packed in $Cr_2O_3$ powder and
refired at 1400 and 1650°C.  In each case a uniform, thin layer
of $Al_2O_3$-$Cr_2O_3$ solid solution formed on the surface.  X-ray diffraction analysis indicated that the outside surface of the layer was
$Cr_2O_3$.  Electron microprobe analyses were used to determine the
composition profiles (Figure 3.4).  The layer formed at 1650°C

---

[*]ALSIMAG 614, 3M Co., Chattanooga, Tenn.

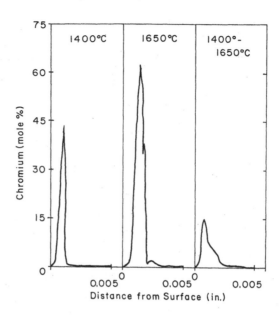

Figure 3.4  Chromium profiles for Al₂O₃-Cr₂O₃ solid solution
surface layers on alumina.  Reprinted with permission
of J. Am. Ceram. Soc. 49 (6) (1965), 330-333.

was much thicker than that formed at 1400°C.  Refiring a specimen
originally treated at 1400°C, at 1650°C without packing again in
$Cr_2O_3$, reduced the maximum percentage of $Cr_2O_3$ and increased the
thickness of the diffused layer.  Little or no grain growth occurred
as a result of the additional heat treatments.

The flexural strengths of bars treated by refiring at 1650°C,
packed in $Cr_2O_3$, are given in Table 3.5.  The treated bars are
8,400 psi stronger than refired controls.  This difference is
significant at the 99.5% level (Student's "t" test).  A smaller
improvement in strength was observed for specimens treated by
packing at 1400°C.

Gruszka, Mistler, and Runk (1970) investigated the effect of
various surface treatments on the flexural strength of high alumina
substrates.  One of these treatments involved evaporation of
chromium onto the surface followed by oxidation in air at elevated
temperatures.  During exposure to the high temperatures most of
the chromium oxide must have evaporated because the surfaces

Table 3.5. Flexural Strength of 96% Alumina Bars with $Al_2O_3$-$Cr_2O_3$ Surface Layers (Bar Dimensions 9/64 x 1/4 x 3 in)

| Treatment | No. Specimens | Flexural Strength Data[*] | | | Standard Deviation psi |
|---|---|---|---|---|---|
| | | Average Flexural Strength psi | Strength Difference psi | Significance Level % | |
| Controls, as cut | 10 | 32,900 | --- | --- | 2.599 |
| Controls, refired at 1650°C, one hour | 19 | 29,100 | --- | --- | 4,048 |
| Packed in $Cr_2O_3$, 1650°C, one hour | 20 | 37,500 | 8,400 | 99.5 | 2,535 |
| Controls, refired at 1650°C, one hour | 9 | 30,200 | --- | --- | 2,600 |
| Controls, surface leached and refired at 1650°C, one hour | 10 | 31,300 | --- | --- | 2,000 |
| Surface leached, packed in $Cr_2O_3$, 1650°C, one hour | 10 | 39,000 | 7,700 | 99.5 | 2,600 |
| Controls, completely leached and refired, 1650°C, one hour | 9 | 28,100 | --- | --- | 1,300 |
| Completely leached, packed in $Cr_2O_3$, 1650°C, one hour | 10 | 31,800 | 3,700 | 99.5 | 2,700 |

[*]Four point loading on a two inch span.

appeared pink rather than the dark green characteristic of high chromium concentrations in $Al_2O_3$.  Substantial improvements in strength were observed compared with refired controls but based upon microscopic examinations which showed less well defined grain boundaries, and the evaporation of chromium oxide, the strengthening effect was attributed to changes in surface texture rather than compressive surface stresses.

Frazier, Jones, Raghavan, McGee, and Bell (1971) confirmed the strengthening effect of $Al_2O_3$-$Cr_2O_3$ solid solution surface layers formed on polycrystalline alumina bodies.  Strength improvements of up to 32% were observed.

Methods for increasing the thickness of the surface layers were developed.  One such method involves leaching of the surface of the ceramic body to form a network of connected pores and then treating it to form the low expansion solid solution surface layer. In 96% alumina bodies, 52% aqueous hydrofluoric acid was used to remove the siliceous intergranular material.  The leaching process was described in more detail by Rishel, Infield, and Kirchner (1968) who used it to enhance the machinability of alumina.

Surface leached and unleached 96% alumina specimens were packed in $Cr_2O_3$ powder and refired at 1650°C (Kirchner, Gruver, and Walker, 1967).  The composition profiles were compared (Figure 3.5) and show that leaching does enhance the penetration of $Cr_2O_3$ into the alumina body.  The areas under the peaks were measured showing that 50% more $Cr_2O_3$ has been transferred to the leached specimen.

The presence of compressive surface stresses was demonstrated by ring tests using rings fabricated especially for the purpose. Treated alumina rings closed 24 μm and 40 μm whereas the controls did not open or close.

The flexural strengths of leached and unleached specimens, treated by packing in $Cr_2O_3$ and refiring were measured (Table 3.5) and show that leached specimens strengthened by packing in $Cr_2O_3$ and refiring are stronger than specimens that are simply leached and refired.

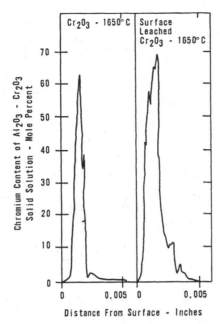

Figure 3.5.   Chromium profiles for $Al_2O_3$-$Cr_2O_3$ solid solution
surface layers on leached and unleached alumina.
Reprinted with permission of J. Am. Ceram. Soc. 50
(4) (1967), 169-173.

The thermal shock resistance of specimens strengthened by
leaching and packing in $Cr_2O_3$ was compared with controls that were
simply leached and refired (Figure 3.6).   The treated specimens
were slightly improved compared with the controls.

### Addition of halides to the packing material

It is well known that additions of halides can be used to
enhance the volatilization of inorganic powders.   In addition,
it was demonstrated earlier that refiring of alumina in environments containing fluorine improves the strength.   Therefore,
alumina specimens were packed in mixtures of $Cr_2O_3$ and additives
containing fluorides and chlorides (Kirchner, Gruver, and Walker,
1968).   Composition profiles for 94% alumina specimens leached and
packed in $Cr_2O_3$ plus various percentages of $CrF_3 \cdot 3.5 H_2O$ are

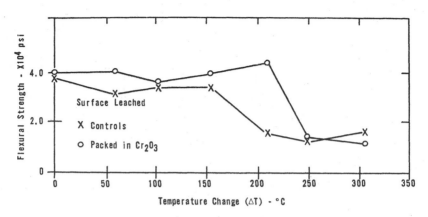

Figure 3.6.   Thermal shock test results for 96% alumina with
              $Al_2O_3$- $Cr_2O_3$ solid solution surface layers.

given in Figure 3.7.   The penetration of $Cr_2O_3$ increases with
increasing fluoride in the mixture.

When $Cr_2O_3$ is used alone as the packing material for
unleached alumina, $SiO_2$ accumulates with $Cr_2O_3$ at the surface.
Leaching and packing in mixtures containing fluorides prevent this
accumulation of $SiO_2$.   Instead, packing in mixtures containing
fluorides leads to accumulation of MgO at the surface.   The MgO
probably reacts to form $MgCr_2O_4$, a low expansion compound with the
spinel structure.   This compound which has a thermal expansion
coefficient of 68 x $10^{-7}$°$C^{-1}$ (25-1000°C), compared with
74 x $10^{-7}$°$C^{-1}$ for $Cr_2O_3$, might aid in formation of compressive
surface stresses.

The appearance of the specimen surface was studied by optical
microscopy.   Unleached alumina treated by packing in $Cr_2O_3$ has a
rather coarse-grained, porous surface.   The presence of fluorine
from leaching or from fluoride in the packing material produces
substantial refinement of the texture of the surface.   $CaF_2$ has
an opposite effect forming very coarse textured surface layers.

As shown by ring tests, compressive surface stresses were
present in the treated specimens.   The forces are greater in
specimens packed in chromium oxide plus chromium fluoride than in
those packed in chromium oxide alone.

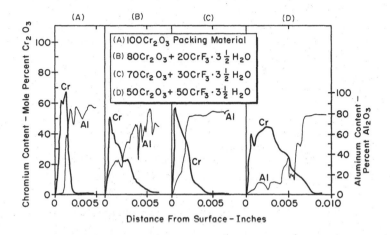

Figure 3.7. Chromium and aluminum profiles for leached 94% alumina packed in $Cr_2O_3$ plus various percentages of $CrF_3 \cdot 3.5\ H_2O$ (refired at 1650°C for one hour). Reprinted with permission of J. Am. Ceram. Soc. 51 (4) (1968), 232.

### 3.2.2 Sapphire

Solid solution surface layers were formed on sapphire by packing in chromium compounds followed by firing at high temperatures (Kirchner, Gruver, and Walker, 1969). Packing in $Cr_2O_3$ powder and refiring at 1500°C for one hour leads to formation of thin $Al_2O_3$-$Cr_2O_3$ surface layers. Refiring at 1500°C for three hours results in thicker layers. Packing in a mixture of 80% $Cr_2O_3$ + 20% $CrF_3 \cdot 3\ 1/2\ H_2O$ and refiring produced even thicker layers.

Slotted rod tests were performed on rods treated by packing in $Cr_2O_3$ and a mixture of 50% $Cr_2O_3$ and 50% $CrF_3 \cdot 3\ 1/2\ H_2O$ with the results shown in Table 3.6.

The flexural strengths of sapphire specimens treated by packing in chromium compounds and refiring were measured and the results are presented in Table 3.7. Improved strength was observed for crystals fired at 1500°C for three hours and then

Table 3.6.  Rod Test Results for Sapphire (Rods 0.1 in diam,
            Slot 1.125 x 0.013 in)

| Treatment | Change in Rod Diameter, in |
|---|---|
| As received control | +0.002 |
| Packed in $Cr_2O_3$, refired 1500°C 3 hours | -0.001 |
| Packed in 50% $Cr_2O_3$ + 50% $CrF_3 \cdot 3 1/2 H_2O$, refired 1500°C one hour | -0.002 |

tested in the 0° and 45° orientations (orientation of the sapphire
crystals was described in Section 2.3.1).  There was no consistent
variation of strength of treated specimens with crystal orientation or layer thickness.

Frazier, Jones, Raghavan, McGee, and Bell (1971) confirmed
that formation of $Al_2O_3$-$Cr_2O_3$ solid solution surface layers on
sapphire rods improved the flexural strength compared with sapphire
rods that were refired according to the same schedule but not
packed in $Cr_2O_3$ containing powders to form the solid solution.
The observed strength increases were 64% and 69%.

Nehring and Jones (1973) studied the flexural creep of surface-
treated sapphire rods.  The rods were packed in a mixture of 80%
$Cr_2O_3$ + 20% $CrCl_3 \cdot 6 H_2O$ and fired at 1500°C for three hours.
The creep rate of the treated rods at 1300°C was much lower than
that of controls but the treated rods failed in brittle fracture
within 30 min after the load was applied whereas the controls had
not fractured after 12 hours.

## Thermal shock resistance of sapphire

The thermal shock resistance of sapphire rods with $Al_2O_3$-$Cr_2O_3$
solid solution surface layers was investigated by Doherty, Tschinkel,
and Copley (1972).  After aging at 1200°C for 100 hours and abrading at room temperature, the specimens were rapidly heated to
1200°C and rapidly cooled in an air stream.  These cycles were

Table 3.7.  Flexural Strength of Sapphire with $Al_2O_3$-$Cr_2O_3$ Surface Layers

| Treatment | No. Specimens | Average Flexural Strength psi | Standard Deviation psi |
|---|---|---|---|
| As received controls, 0° orientation | 5 | 88,800 | 25,200 |
| Packed in $Cr_2O_3$, 1500°C, 3 hours 0° orientation | 2 | 220,200 | 35,600 |
| Packed in $Cr_2O_3$, 1500°C 3 hours 45° orientation | 2 | 227,600 | |
| Packed in $Cr_2O_3$, 1500°C, one hour not oriented | 2 | 154,300 | --- |
| Packed in 50% $Cr_2O_3$ + 50% $CrF_3$ · 3 1/2 $H_2O$ 1500°C, one hour, 0° orientation | 4 | 109,300 | 19,800 |

repeated with increasing air mass flux until failure occurred.
The temperature and stress at failure were calculated.  The
average fracture stress in the thermal shock test was slightly
greater for specimens that were treated and abraded compared with
controls that were simply abraded.

### 3.2.3  Polycrystalline Titania

The thermal expansion of $TiO_2$-$SnO_2$ solid solutions is given
in Figure 3.2 and the calculated compressive surface stresses for
various solid solution compositions are given in Figure 3.3.
Based on this information it was decided to form $TiO_2$-$SnO_2$ solid
solution surface layers on polycrystalline titania bodies.

In contrast to the experiments with polycrystalline alumina
in which fired specimens were treated, the titania specimens were
treated in the unfired or "green" condition and then fired to
mature the body and react the surface layer material in one step
(Kirchner and Gruver, 1966).  $TiO_2^*$ was mixed with 1/4% $WO_3$ added
as a grain growth inhibitor and appropriate binders.  The mixture
was granulated by forcing it through a 50-mesh sieve.  Then, it
was pressed at 10,000 psi to form bars 3 x 3/8 x 1/4 in.  The
following methods were used to form $TiO_2$-$SnO_2$ surface layers.

1.  $SnCl_4$ + $H_2SO_4$ solution (1 gram $SnCl_4$ · $5H_2O$ + 0.4 ml $H_2SO_4$
    conc diluted to one ml) was applied to the surfaces by dipping.
2.  Organic compounds of tin (tri-n-butyl tin oxide, tri-butyl-tin
    methacrylate) were applied to the surfaces by pouring.
3.  $SnO_2$ slip painted on the surface.
4.  $SnO_2$ powder used as a packing material during firing.

Dipping in $SnCl_4$ + $H_2SO_4$ solution was the most successful
method.  As the techniques for preparing the batch and dry pressing
the specimens improved, it became more and more difficult to
impregnate the specimens with the solution.  To increase the

---

*Heavy grade, TAMCO Niagara Falls, New York.

porosity of the surface before dipping, the binder was burned out
of some of the specimens by heating to 700°C.  Then, the specimens
were fired in an electric furnace for one hour at 1400°C.

The tin content at various distances from the surface was
measured by electron microprobe and by x-ray diffraction analysis.
The tin oxide contents ranged up to 8% at the surface and decreased
to 0% at a depth of 0.030 to 0.040 in.  The fired control specimens
contained a trace of $SnO_2$ as a result of reaction with the furnace
atmosphere.

The presence and relative magnitudes of the surface forces
were determined by ring tests using rings 1 3/8 in OD with 1/8 in
wall thickness and 5/8 in high.  Only the outside surface of each
ring was treated.  The surfaces of the control rings had small
compressive forces causing slight closing of the rings.  These
forces were caused by small amounts of $SnO_2$ transferred through
the furnace atmosphere to the controls.  Rings treated with the
$SnCl_4 + H_2SO_4$ solution closed about 11 times as much, indicating
much greater surface forces.

The strengths of the treated specimens were greater than the
controls in all but a few experiments.  In some cases substantial
decreases in the standard deviations were observed indicating
improved uniformity and reliability.  The greatest increase in
strength was obtained for specimens treated with the $SnCl_4 + H_2SO_4$
solution after binder burn out.  In this case the average strength
increased 5000 psi, from 19,400 psi to 24,400 psi.

### 3.2.4  Polycrystalline Spinel ($MgAl_2O_4$)

Solid solutions, in which part of the alumina in spinel is
replaced by chromia, have lower expansion coefficients than the
pure spinel (Figure 3.8).  Therefore, polycrystalline spinel bodies
were strengthened by formation of $MgAl_2O_4$ - $MgCr_2O_4$ solid solution
surface layers (Kirchner, Gruver, and Walker, 1967).

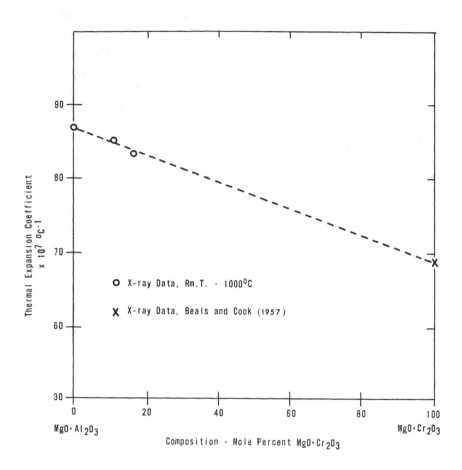

Figure 3.8.   Linear thermal expansion of MgO · Al$_2$O$_3$ - MgO · Cr$_2$O$_3$
              solid solutions.   Reprinted with permission of J. Am.
              Ceram. Soc. 50 (4) (1967), 169-173.

Two spinel bodies were treated including a commercial body[*]
supplied as bars approximately 3 x 0.34 x 0.135 in, and as hollow
cylinders approximately 0.245 in OD x 0.120 in ID x 2.25 in long
and an experimental body prepared by methods similar to those of
Ryshkewitch (1960).   To increase the thickness of the surface

---

[*]DEGUSSA SP-23, Degussa Incorporated, Kearney, N.J.

layers the specimens were leached with 52% aqueous hydrofluoric
acid before treating to form the low expansion solid solution
surface layer.  The penetration of the chromium increased with
increasing leaching time as shown in Figure 3.9 for the experimental spinel body.  Similar increases in chromium content and
depth of penetration with increasing leaching time were observed
for DEGUSSA SP-23 spinel.

Because only chromium oxide, rather than $MgCr_2O_4$ is added
to the surface, the surface was expected to be deficient in MgO.
Composition profiles were determined and are given in Figure 3.10.
In the untreated specimen, the magnesium content varies but
averages about 30 mole %.  In the specimen with the $MgCr_2O_3$-$MgAl_2O_4$
solid solution surface layer, the magnesium content increases from
about 10 mole % near the surface to about 25% at a depth of 0.010
in confirming the original expectations.  The ratio of divalent
to trivalent cations is important in determining the thermal expansion coefficients of spinel.  Kirchner and Gruver (1965) showed

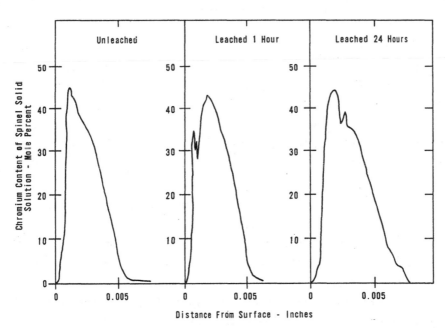

Figure 3.9.   Chromium profiles for experimental spinel body packed
in $Cr_2O_3$ and fired at 1650°C.  Reprinted with permission
of J. Am. Ceram. Soc. 50 (4) (1967), 169-173.

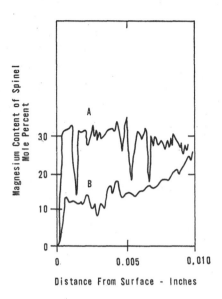

Figure 3.10.   Magnesium profiles for experimental spinel body
               fired at 1650°C.   (A) Spinel with no chromium solid
               solution layer.   (B) Spinel with chromium solid
               solution layer.   Reprinted with permission of J. Am.
               Ceram. Soc. 50 (4) (1967), 169-173.

that there is a substantial decrease in thermal expansion coefficient with increasing $Al_2O_3$ content of magnesium aluminum spinels.
Presumably the same is true for magnesium chromium spinels and
their solid solutions with magnesium aluminum spinels.  Therefore,
one would expect compressive stresses to be increased as a result
of this factor in addition to the increases expected as a result
of the reduction in thermal expansion coefficients indicated for
supposedly stoichiometric compositions in Figure 3.8.

     The existence of the compressive surface stresses in the
treated specimens was demonstrated by ring tests.  Rings of
DEGUSSA SP-23 spinel, packed in $Cr_2O_3$ and refired at 1750°C for
one hour were closed 24 µm and 8 µm after cutting, confirming the
presence of the compressive surface stresses.

     The flexural strengths of both types of spinel were increased
by the treatments.  Refiring temperatures ranging from 1400 to
1750°C were effective.  Leaching for short periods, followed by

packing and refiring resulted in additional improvements in
strength as shown in Figure 3.11 for the experimental spinel body.
Similar results were observed for DEGUSSA SP-23 spinel except that
its strength was not degraded by prolonged leaching.  This difference may have occurred because the DEGUSSA SP-23 body is less
porous than the experimental spinel so that it is less easily
penetrated by the hydrofluoric acid.

Hollow cylinders of DEGUSSA SP-23 spinel were thermally shocked
by quenching in water from various temperatures.  The remaining
flexural strengths were measured by three point loading on a one
inch span with the results shown in Figure 3.12.  Because the span
to depth ratio was small, the measured strengths are too high.
Neverthlesss, the results are useful for comparison and show that
the chemically strengthened specimens are stronger and more thermal
shock resistant than the controls.

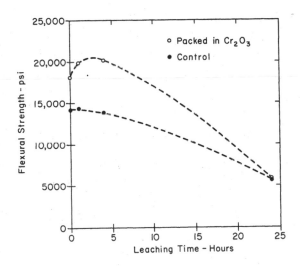

Figure 3.11.  Flexural strengths of leached and chemically strengthened experimental spinel body fired at 1750°C for one
hour.  Reprinted with permission of J. Am. Ceram.
Soc. 50 (4) (1967), 169-173.

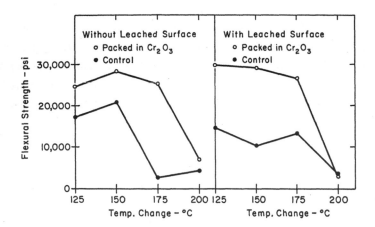

Figure 3.12.  Thermal shock test results for DEGUSSA spinel fired
              at 1750°C for one hour.  Reprinted with permission
              of J. Am. Ceram. Soc. 50 (4) (1967), 169-173.

### 3.2.5  Polycrystalline Magnesia

The thermal expansion of MgO-NiO solid solutions decreases
with increasing NiO content as shown in Table 3.8 (Kirchner and
Gruver, 1965).  Therefore, MgO-NiO solid solution surface layers
on magnesia bodies are subjected to compressive stresses on cooling
with resulting improvement in strength.

Table 3.8.  Thermal Expansion of MgO-NiO Solid Solutions
            (25-1000°C)

| Composition (mole fraction) | Thermal Expansion Coefficient $\times\ 10^{7}$°C$^{-1}$ |
|---|---|
| MgO | 139 |
| 0.8 MgO-0.2 NiO | 137 |
| 0.6 MgO-0.4 NiO | 134 |
| 0.4 MgO-0.6 NiO | 133 |

Additions of CoO have a similar effect in reducing the thermal expansion of MgO. $Li^+$ and $Cr^{3+}$ can also be combined to reduce the expansion coefficient.

Three types of magnesia were used in strengthening experiments (1) a commercial body purchased from Honeywell, Inc. in the form of rectangular bars  (2) an experimental body prepared by isostatic pressing of 3/8 x 1/4 x 3 in bars of magnesia powder, freshly prepared by decomposing $MgCO_3$, and sintering at 1320°C for 40 hours as described by Harrison (1963) and  (3) a hot pressed body from which small rectangular bars were cut.

The commercial magnesia bars were packed in various powders and refired at 1400°C for two hours. Refiring degraded the strengths of the controls. Packing and refiring seemed to protect the specimens from degradation. In the absence of rod test results and other evidence it is uncertain whether or not compressive surface stresses were induced and, if so, whether or not the stresses had a role in preventing further strength degradation.

The MgO-NiO surface layer was very light green in color. Removal of less than 0.001 in of material by orthophosphoric acid removed the green color completely. Therefore, the surface layer is very thin. Analysis of x-ray diffraction patterns showed little peak shifting relative to the untreated material. The light color and the small amount of peak shifting indicates that the NiO content of the surface layer was small.

The experimental magnesia bars were packed in CoO and refired at 1400°C for 14 min. A dark red surface layer was formed. The x-ray diffraction patterns of these surfaces, using copper radiation, were unsatisfactory because of high background radiation. The treated specimens were slightly stronger than the controls as shown in Table 3.9.

In other experiments 0.5 MgO-0.5 NiO solid solution surface layers were formed on hot pressed MgO during the hot pressing treatment. Bars were cut from the coated discs and the flexural strengths were measured by three point loading on a one half inch span. The average strength was 39,400 psi compared with 36,100 psi for the controls, indicating a small improvement.

Table 3.9.  Flexural Strength of Chemically Strengthened Experimental Magnesia

| Treatment | No. Specimens | Flexural Strength Data* Average Flexural Strength psi | Strength Difference psi |
|---|---|---|---|
| Controls, as fabricated | 4 | 18,600 | --- |
| Controls, polished, refired 1400°C, 15 min | 5 | 17,600 | -1,000 |
| Controls, packed in CoO, 1400°C, 15 min | 5 | 21,300 | +2,700 |

*Four point loading on a two inch span.

Cutler and Brown (1964) made a limited study of the effect
of doping the surfaces of polycrystalline magnesia ceramics with
$Fe_2O_5$ and NiO. The test bars were treated as follows:

1. $Fe_2O_3$ - Polycrystalline magnesia specimens were dipped in
   ferric chloride solution and refired at 1080°C for 25 minutes.
   The diffusion layer was 1-10 μm thick.
2. NiO - Polycrystalline magnesia specimens were packed in NiO
   powder and refired at temperatures in excess of 1500°C for
   90-110 minutes.

Flexural strength measurements yielded a strength increase from
11,800 psi to 16,400 psi (38%) for the specimens doped with $Fe_2O_3$.
Small increases (3-5%) that probably were not statistically significant were observed for the specimens doped with NiO. In contrast
with these results for $Fe_2O_3$, Kirchner and Gruver (1966) found
that the strength of the stronger commercial magnesia was severely
degraded by packing in $Fe_2O_3$ and refiring at 1400°C for two hours.
Analysis of peaks in the back reflection region of the x-ray diffraction pattern showed little or no shifting of peak positions,
providing preliminary evidence that little or no solid solution
formation occurred. On the other hand, new peaks appeared in the
pattern indicating the presence of a second phase which might be
responsible for the weakening.

Plastic behavior has a substantial role in the fracture of
pure magnesia. Liu, Stokes and Li (1964) showed that addition
of small amounts of NiO or MnO in solution in the body increased
the compressive yield strength of MgO nearly four-fold. Solid
solution formation in the surfaces of magnesia bodies can be
expected to increase the resistance of the material to dislocation
motion. Because there are alternative explanations for the observed
strengthening of magnesia one cannot, without additional evidence,
assert that one or another of the proposed mechanisms is responsible
for the observed increases in strength.

## 3.2.6  Periclase (MgO)

Bickelhaupt, Cox, and Hurst (1967) and Bickelhaupt (1968,1970)
investigated the effect of compressive surface stresses on the
strength of periclase (magnesium oxide) single crystals.  These
surface layers were formed by diffusing NiO or MnO into the
surfaces of rectangular bars of MgO at high temperatures.  MgO-
NiO and MgO-MnO solid solutions have lower thermal expansion coefficients than MgO so that, during cooling to room temperature,
compressive surface stresses are induced because of the greater
thermal contraction of the underlying material.  Based upon measured
thermal expansion coefficients and certain assumptions involving
Young's modulus and creep at elevated temperatures, the stress
profile shown in Figure 3.13 was calculated.  This profile shows a
maximum compressive stress of about 14,000 psi in the surface and
relatively low tensile stresses in the interior.

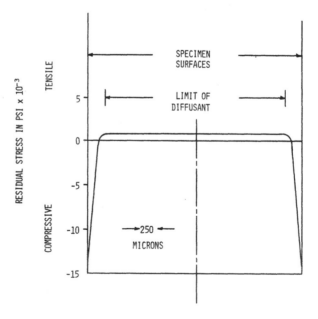

Figure 3.13.   Calculated residual stress profile for MgO-NiO surface
              layers on periclase single crystals.

The yield strength of specimens with solid solution surface layers was approximately 7000 psi higher than that of specimens that were simply thermally annealed. Since some of the surface layer was removed by abrading prior to testing, the correlation between the calculated surface stress and the strength increase was considered satisfactory. Microscopic evidence of arrest of a crack by the solid solution surface layer was presented (Bickelhaupt, 1968).

### 3.2.7 Polycrystalline Nickel Oxide

Nickel oxide has a very high thermal expansion coefficient ($143 \times 10^{-7} \,^{\circ}C^{-1}$, 25-936°C). As indicated in Section 3.25, solid solutions in the system MgO-NiO have lower expansion coefficients. Therefore, solid solution surface layers of these compositions, formed on nickel oxide specimens, were expected to lead to compressive surface stresses and improved strengths.

Strong nickel oxide bodies have been made by hot pressing (Spriggs, Brissette, and Vasilos, 1964 and Harrison, 1965). Discs, 7/8 or 1 1/8 in diam, of nickel oxide were hot pressed by similar methods and cut to form rectangular bars (Kirchner, Gruver and Walker, 1966 and Kirchner, Gruver, Platts, and Walker, 1967). Multilayer discs consisting of 0.60 NiO + 0.40 MgO surface layers and NiO main body were also prepared. Solid solution surface layers formed by reaction of oxide powders on NiO yielded an average flexural strength of 16,500 psi compared with NiO controls which averaged 10,900 psi. Solid solution surface layers formed by reaction of NiO with $Mg(OH)_2$ yielded an average flexural strength of 16,900 psi compared with 13,300 psi for controls. The results of these experiments show reasonable evidence of strengthening but in the absence of rod test results and other evidence the strengthening mechanism is uncertain.

## 3.2.8  Silicon Nitride

Skrovanek and Bradt (1976) used solid solutions of alumina
in silicon nitride to strengthen reaction bonded silicon nitride.
These Si-Al-O-N solid solutions are often considered to have lower
thermal expansion coefficients than silicon nitride so that compressive surface stresses might be expected on cooling to room
temperature.  Alumina was deposited on the surfaces of silicon
nitride specimens by decomposition of aluminum sulfate.  During
subsequent heat treatments the alumina was diffused into the specimens to form the solid solution surface layers which were about
225 μm thick in one case.  The flexural strengths of specimens
with several different heat treatments were measured.  The highest
average strength, 53,300 psi, was observed for specimens heat
treated at 1390°C, just below the melting point of silicon.  The
average strength of as-received controls was 35,300 psi so the
treatment resulted in a 51% increase in strength.  A shift in
fracture origins from the surface to the interior is consistent
with strengthening by compressive surface stresses.

## 3.3  LOW EXPANSION COMPOUND SURFACE LAYERS FORMED BY REACTION WITH THE SURFACE

Low expansion compound surface layers were formed by processes
involving chemical reactions with the body and by chemical vapor
deposition in which the low expansion phase simply coated the body.
Some examples of the first case are described in this section and
the effect of coatings formed by chemical vapor deposition is
described in the next section.

### 3.3.1  Polycrystalline Alumina

Experience has shown that, in order to obtain substantial
increases in strength of a ceramic body by chemical reactions to
induce compressive surface stresses, the reactions must be uniform

over the surface. Such uniformity is best obtained by reactions
between the body and a volatile material rather than by solid state
reaction and diffusion from point contacts with powders. Therefore,
reactants having sufficient vapor pressure at reasonable temperatures were chosen.

The thermal expansion coefficients of ceramics with complex
formulae and open crystal structures tend to be lower than that of
alumina (Kirchner, Scheetz, Brown, and Smyth, 1962). Therefore,
after cooling, surface layers of these materials will be in compression.

Because alumina (corundum) has a very densely packed crystal
structure, the products of most chemical reactions involving alumina
will have a larger volume than the reactants. If the lattice
framework is not disrupted by the reaction but tends to expand
to form the new phase, the reaction itself may induce stresses
in the surfaces. The volume changes for several reactions were
calculated with the following typical results:

$$CaO + 2Al_2O_3 \rightarrow CaAl_4O_7 \qquad\qquad \Delta V = 31\%$$

$$CaO + 6Al_2O_3 \quad CaAl_{12}O_{19} \qquad\qquad \Delta V = 15\%$$

$$3Al_2O_3 + 2SiO_2 \quad Al_6Si_2O_{13} \text{ (mullite)} \qquad \Delta V = 9.7\%$$

Based on these calculations it was decided to form calcium aluminate,
mullite, and spinel surface layers on alumina bodies.

Two alumina bodies were selected for treatment. One was
a 99.9% $Al_2O_3$ body[*] in the form of extruded cylindrical rods,
0.118 in diam and 2.25 in long, with a 3 μm average grain size,
a density of 3.90-3.93 $g/cm^3$, and a surface finish better than
20 μin rms. The other body was the 96% $Al_2O_3$ body[+] in the form
of extruded cylindrical rods, 0.125 in diam.

---

[*]AD999 Coors Porcelain Co., Golden, Colo.

[+]ALSIMAG 614, 3M Co., Chattanooga, Tenn.

The rods were packed in powdered reactants and refired to form the surface layers.  $CaCO_3$ and calcium aluminate were used to form calcium aluminate surface layers.  The calcium aluminate packing materials were less satisfactory than $CaCO_3$ because of lower volatility.  Mullite surface layers were a special problem because of the low volatility of $SiO_2$ which would otherwise have been used as a packing material.  Silicon carbide was found to be a satisfactory packing material.  Apparently, the silicon carbide tends to oxidize, forming silicon monoxide which, at the firing temperatures used, is volatile.  The silicon monoxide then reacts with the alumina to form mullite.

## Calcium aluminate surface layers

Composition profiles for 96% $Al_2O_3$ and 99.9% $Al_2O_3$ rods packed in $CaCO_3$ at 1350°C, are given in Figure 3.14.  The diameters of the specimens increased by as much as 0.006 in.  Despite the longer time at 1350°C for the 99.9% $Al_2O_3$ (3 hours) less CaO was deposited on its surface than on the 96% $Al_2O_3$.  This difference may be the result of lower reactivity of the 99.9% $Al_2O_3$ which may lead to reevaporation.

Kohatsu (1967) studied solid state reactions between CaO and $Al_2O_3$ and, in one of his experiments, CaO was evaporated onto the surface of a polycrystalline alumina pellet.  X-ray diffraction analysis confirmed the formation of $CaAl_4O_7$ and $CaAl_{12}O_{19}$ and showed the absence of preferred orientation of these compounds.  In the present experiments which were done at lower temperatures and shorter times than those employed by Kohatsu, x-ray diffraction analysis indicated the presence of complex reaction products.  It is likely that the entire series of calcium aluminates form on these bodies.  Because the thermal expansion coefficient of $CaAl_4O_7$ (45 x $10^{-7}$°$C^{-1}$, 100-1200°C) (Rigby and Green, 1943) is much lower than those of the alumina bodies, any strengthening may result either from compressive stresses caused by volume increase during chemical reaction or from differences in thermal expansion.

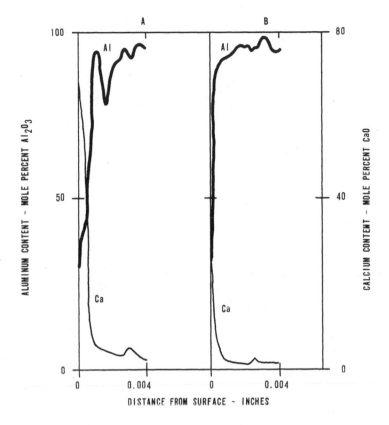

Figure 3.14.   Aluminum and calcium profiles for alumina packed in
               CaCO₃.  (A) 96% alumina, 1350°C, one hour.  (B)
               99.9% alumina, 1350°C, three hours.  Reprinted with
               permission of Trans. Brit. Ceram. Soc. 70 (6) (1971),
               215-219.

Rod tests of 96% $Al_2O_3$ specimens, packed in $CaCO_3$ for four
hours showed the presence of compressive surface stresses.  Treated
rods closed 0.0037 in, whereas as received controls and refired
controls opened.

The flexural strengths of the 99.9% $Al_2O_3$ specimens, strengthened by packing in $CaCO_3$ are reported in Table 3.10.  The results
are presented in three groups depending on the method used to
polish the rods before treatment.  Substantial increases in strength
were observed for treatments at temperatures ranging from 1250-1350°C.

Table 3.10. Flexural Strength of 99.9% $Al_2O_3$ Strengthened by Packing in $CaCO_3$

| Treatment | No. Specimens | Flexural Strength Data* Average Flexural Strength psi | Strength Difference psi |
|---|---|---|---|
| **Specimens not polished** | | | |
| Controls, as received | 5 | 56,300 | --- |
| Packed in $CaCO_3$, 1250°C, 3 hours | 3 | 93,800[+] | +37,500 |
| Packed in $CaCO_3$, 1300°C, 3 hours | 3 | 73,300[+] | +17,000 |
| Packed in $CaCO_3$, 1350°C, 3 hours | 3 | 82,000 | +27,700 |
| Packed in $CaCO_3$, 1350°C, 3 hours | 3 | 85,400[ƒ] | +29,100 |
| Packed in $CaCO_3$, 1400°C, 3 hours | 3 | 58,600[#] | +2,300 |
| **Polished using 400 and 600 grit SiC paper and 15 μm diamond paste** | | | |
| 1250°C, 3 hours | 4 | 58,100[x] | +1,800 |
| Packed in $CaCO_3$, 1250°C, 3 hours | 4 | 83,800[x] | +27,500 |
| 1350°C, 3 hours | 4 | 67,300[x] | +11,000 |
| Packed in $CaCO_3$, 1350°C, 3 hours | 4 | 75,000[x] | +18,700 |
| **Polished using 15 μm and 6 μm diamond paste** | | | |
| 1250°C, 3 hours | 6 | 77,700[x] | +21,400 |
| Packed in $CaCO_3$, 1250°C, 3 hours | 6 | 79,400[x] | +23,100 |
| Packed in $CaCO_3$, 1250°C, 3 hours, dipped in silicone oil (12500 cSt) | 6 | 93,100[x] | +36,800 |

*Four point loading on a two inch span unless otherwise noted.
[+]One low value omitted from average.
[ƒ]The highest value was 121,300 psi.
[#]Only one specimen was tested. The other two specimens were fluxed by $CaCO_3$.
[x]Four point loading on a one inch span at about 20% relative humidity.

At higher temperatures the onset of the fluxing reaction prevented
strengthening.  The highest individual strength value, 121,300 psi,
was observed for an unpolished specimen, packed in $CaCO_3$ and
refired at 1350°C for three hours.

The variability of the early results with the 99.9% $Al_2O_3$
specimens was tentatively attributed to variations in surface
condition prior to treatment.  Therefore, in later experiments
the original surface was removed from the specimens by polishing
before the specimens were treated.  The first polishing treatment
involved 400 and 600 grit SiC paper followed by 15 μm diamond
paste.  Higher strengths were observed for refired specinens.  In
an attempt to obtain additional improvements, the specimens were
polished using 15 μm and 6 μm diamond paste.  The refired controls
were much stronger than previously, perhaps because of the improved
surface.  The treated specimens were slightly stronger than the
refired controls.  Due to speculation that unreacted CaO was
reacting with water vapor and enhancing stress corrosion during
testing, one group of specimens was dipped in silicone oil to
reduce stress corrosion.  Normally, an increase of about 6,000
psi should be expected as a result of this treatment.  Instead,
an increase of 14,000 psi was observed, the larger increase being
consistent with the above proposed mechanism of stress corrosion
involving CaO.

To provide additional evidence that the strengths of alumina
specimens treated by packing in $CaCO_3$ were prevented from achieving
their highest strengths by stress corrosion, hot pressed alumina
specimens were treated by packing and refiring at 1350°C for three
hours.  The specimens protected by silicone oil had an average
flexural strength of 109,100 psi which is 24,700 psi (Table 3.11)
higher than those not so protected.  Because this increase is
greater than normally expected, it represents additional evidence
of an unusual stress corrosion mechanism.

The flexural strengths of 96% $Al_2O_3$ specimens treated by
packing in $CaCO_3$ were consistently increased by up to 40%.  The
strongest group was packed in $CaCO_3$ and refired at 1350°C for four

Table 3.11.  Flexural Strength of Hot Pressed Alumina[*] Treated by Packing in $CaCO_3$ (Polished[+] Rods 0.1 in diam, Refired at 1350°C for 3 hours)

| Specimen No. | As Polished | Flexural Strength[†] - psi Packed in $CaCO_3$ | Packed in $CaCO_3$ Dipped in Silicone Oil |
|---|---|---|---|
| 1 | 85,000 | 102,700 | 120,400 |
| 2 | 82,400 | 82,400 | 118,200 |
| 3 | 63,100 | 68,200 | 99,000 |
| 4 | - - - | - - - | 98,700 |
| Average | 76,900 | 84,400 | 109,100 |

[*] AVCO Corp., Lowell, Mass.

[+] Polished using SiC paper of various grits and 15 µm diamond paste.

[†] Four point loading on a one inch span.

hours.   The average strength was 69,200 psi compared with 49,700
psi for controls.   The poor results observed at higher treatment
temperatures may be the result of relief of stresses by plastic
deformation or of the onset of the fluxing reaction.

### Mullite surface layers

Composition profiles for 99.9% $Al_2O_3$ specimens packed in SiC
and fired at 1500°C are given in Figure 3.15.   Substantial $SiO_2$
was present in the surface.   The presence of mullite was verified
by x-ray diffraction analysis.   Mullite has a thermal expansion
coefficient of 53 x $10^{-7}°C^{-1}$ (20-1100°C) (Geller and Insley, 1932)
so that, on cooling, compressive surface stresses are induced in
the mullite.

The flexural strengths of 99.9% $Al_2O_3$ specimens strengthened
by packing in SiC are given in Table 3.12.   Increases in strength
were observed at all treatment temperatures.   These increases are
greater than would be expected simply as a result of refiring as
can be seen by comparison with refired controls in Table 3.10.
The highest average strength, 84,700 psi, was observed for specimens packed in SiC and refired at 1300°C.

The 96% $Al_2O_3$ specimens were also strengthened by packing in
SiC.   Rod tests of specimens packed at 1500°C for one hour showed
no deflection whereas control specimens opened 0.002 in.   Assuming
that the opening observed for the control specimens was caused by
damage during slotting and that this deflection was counteracted
by compressive surface stresses in the treated specimens, these
observations are tentative evidence of compressive surface stresses
in the treated specimens.   The flexural strengths are given in
Table 3.13.   The results show that the treatments were less
effective at 1500°C than at 1300 and 1400°C.

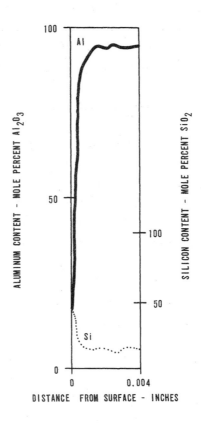

Figure 3.15.  Aluminum and silicon profiles for 99.9% alumina
              packed in SiC and fired at 1500°C.  Reprinted with
              permission of Trans. Brit. Ceram. Soc., 70 (6)
              (1971), 215-219.

## Other surface treatments

In other experiments, surface layers of Li-Al spinel, Ni-Al
spinel, and Mg-Al spinel were formed on 96% $Al_2O_3$ specimens.  Small
increases in strength were consistently observed for specimens
with Li-Al spinel surface layers.  Little or no strengthening was
observed for the other treatments.

Frazier, Jones, Raghavan, McGee, and Bell (1971) formed
$CoAl_2O_4$ (spinel) surface layers on a 99% alumina body by packing

Table 3.12. Flexural Strength of 99.9% $Al_2O_3$ Strengthened by Packing in SiC

| Treatment | No. Specimens | Flexural Strength Data* Average Flexural Strength psi | Strength Difference psi |
|---|---|---|---|
| Controls, as received | 5 | 56,300 | - - - |
| Packed in SiC, 1250°C, 3 hours | 3 | 75,700 | +19,400 |
| Packed in SiC, 1300°C, 3 hours | 3 | 84,700 | +28,400 |
| Packed in SiC, 1350°C, 3 hours | 3 | 77,500 | +21,200 |
| Packed in SiC, 1400°C, 3 hours | 3 | 59,000 | + 2,700 |
| Packed in SiC, 1500°C, 3 hours | 3 | 78,200[+] | +21,900 |

*Four point loading on a two inch span.
[+]Highest strength 106,000 psi.

Table 3.13.  Flexural Strength of 96% $Al_2O_3$ Strengthened by Packing in SiC

| Treatment | No. Specimens | Flexural Strength Data* Average Flexural Strength psi | Strength Difference psi |
|---|---|---|---|
| Controls, as received | 19 | 49,700 | - - - |
| Controls, refired 1300°C, 19 hours | 5 | 53,300 | + 3,600 |
| Packed in SiC, 1300°C, 19 hours | 5 | 63,100 | +13,400 |
| Controls, refired 1400°C, 4 hours | 5 | 55,400 | + 5,700 |
| Packed in SiC, 1400°C, 4 hours | 5 | 64,900 | +15,200 |
| Controls, refired 1500°C, 1 hour | 5 | 62,500 | +12,800 |
| Packed in SiC, 1500°C, 1 hour | 5 | 58,100 | + 8,400 |

*Four point loading on a two inch span.

the specimens in $Co_xO_y$ and refiring at 1350°C for 1 1/2 hours. The presence of the spinel was determined by x-ray diffraction analysis. Electron microprobe measurements indicated cobalt penetration of 10 μm. No evidence of compressive surface stresses was presented. The average flexural strength of the specimens increased by 18% to 48,900 psi as a result of the treatments.

### 3.3.2 Sapphire

The compound surface layers that strengthened polycrystalline alumina bodies were expected to strengthen sapphire crystals. Therefore, sapphire rods were packed in $CaCO_3$ and SiC and refired to form calcium aluminate and mullite surface layers.

Composition profiles for sapphire treated by packing in $CaCO_3$ showed substantial deposition of CaO on the sapphire, but somewhat less diffusion than in polycrystalline alumina. This deposition increased the rod diameter by 0.001 in. X-ray diffraction patterns of the surface material are consistent with formation of calcium aluminates but are not good enough to identify particular phases. Composition profiles for sapphire packed in SiC were not determined but are expected to be similar to Figure 3.15. X-ray diffraction patterns of the surfaces of sapphire specimens packed in SiC showed mullite formation.

Rod tests were used to indicate the presence of compressive surface stresses in the treated sapphire specimens with the results shown in Table 3.14. The control specimens increased in diameter by as much as 0.003 in when slotted. As explained previously these increases may be the result of residual tensile stresses in the surfaces or of damage to the surfaces of the slot. In some cases this damage was visible without the aid of magnification and appeared to consist of groups of cracks. The diameter of a rod packed in $CaCO_3$ showed little decrease after slotting. However, when a second slot was made so that the thickness of the sapphire was a smaller proportion of the total thickness, a marked decrease in diameter was observed. The evidence presented in Table 3.14,

Table 3.14.  Rod Test Results for Sapphire Packed in CaCO₃ and SiC (Rods 0.1 in diam, Slot 1.125 x 0.013 in)

| Treatment | Change in Rod Diameter, in |
|---|---|
| Control, as received | +0.002 |
| Packed in CaCO₃, 1350°C, 3 hours | -0.005 |
| Packed in SiC, 1350°C, 8 hours | +0.001 |

plus the evidence assembled in Section 3.3.1 as a result of experiments with polycrystalline alumina, indicates that compressive stresses were present in the surfaces of the treated specimens. This evidence is less conclusive than desired because a large fraction of the crystals break during slotting. The crystals may break because of the low $K_{IC}$ values or for some reason related to the residual stress. Also, the reaction layers on sapphire tend to be thinner than those observed for polycrystalline ceramics. Therefore, there is less than the usual deflection and it is less certain that significant differences in deflection are observed.

The flexural strengths of specimens with calcium aluminate and mullite surface layers are given in Table 3.15. For calcium aluminate surface layers the best results, 269,800 psi, were obtained by packing in $CaCO_3$ and refiring at 1350°C for three hours. For mullite surface layers, good results were obtained by packing in 600-mesh SiC and refiring at 1350°C (8 hours), 1400°C (4 hours) and 1500°C (6 hours). Because of the difficulty involved in removing the crystals from the partially sintered silicon carbide packing material, these results probably represent the strengths of abraded specimens.

## Thermal shock resistance of sapphire

The thermal shock resistance of sapphire rods treated to form calcium aluminate and mullite surface layers was investigated by Doherty, Tschinkel, and Copley (1972). The sapphire rods were aged for 100 hours at 1200°C and abraded at room temperature. Then, the specimens were thermally shocked by rapidly heating to 1200°C and rapidly cooling in an air stream. These cycles were repeated with increasing air mass flux until failure occurred. The temperature and stress at failure were calculated. The experiments showed that the thermal shock resistance of both types of treated rods was improved in comparison with that of as-received controls that were not abraded. The improvement in thermal shock resistance was especially great for the specimens with mullite surface layers. The authors detected the presence of cristobalite

Table 3.15. Flexural Strength of Sapphire Strengthened by Compound Surface Layers

| | No. Specimens | Flexural Strength Data* | |
| --- | --- | --- | --- |
| | | Average Flexural Strength psi | Standard Deviation psi |
| Calcium aluminate | | | |
| Controls, as received, 90° orientation | 5 | 104,900 | 23,900 |
| Packed in CaCO₃, 1350°C, 3 hours 90° orientation | 2 | 269,800 | --- |
| Packed in CaCO₃, 1350°C, 8 hours 0° orientation | 4 | 168,600 | 52,800 |
| Suspended over CaCO₃, 1350°C, 8 hours 0° orientation | 4 | 141,500 | 17,200 |
| Mullite+ | | | |
| Controls, as received (selected for best quality) | 3ᶠ | 161,400 | 68,800 |
| Packed in SiC#, 1350°C, 8 hours | 4ᶠ | 236,900 | 24,300 |
| Suspended over SiC, 1350°C, 8 hours | 4ᶠ | 176,600 | 38,200 |
| Packed in SiC, 1400°C, 4 hours | 4ᶠ | 257,300 | 10,000 |
| Packed in SiC, 1500°C, 6 hours | 4ᶠ | 258,300 | 23,100 |
| Packed in SiC, 1650°C, 1 hour | 4 | 130,400 | 57,900 |
| Suspended over SiC, 1650°C, 1 hour | 4 | 183,200 | 20,700 |

*Three point loading on a one inch span.
+All specimens tested in 0° orientation.
#600-mesh SiC.
ᶠOne low value omitted from average.

in the surface layers after aging and proposed that the unexpected
magnitude of the improvement was the result of this cristobalite
formation.

### 3.3.3 Polycrystalline Magnesia

Rhodes, Sellers, Vasilos, Heuer, Duff, and Burnett (1966)
accidentally formed forsterite ($Mg_2SiO_4$) surface layers on hot
pressed magnesia during vacuum annealing treatments at 1100°C in
a furnace with fibrous silica insulation.  The surface layers
were about 2 μm thick and appeared to be poorly bonded to the MgO
specimens.  Forsterite has a lower thermal expansion than magnesia
(94 x $10^{-7}$/°C vs 120 x $10^{-7}$/°C for the temperature range 100-200°C).
Therefore, after cooling to room temperature the surface layer was
in compression.  The flexural strengths of these specimens ranged
up to 55,000 psi which were the highest values measured by these
investigators for any magnesia specimens.  The fracture stress of
annealed control specimens was 32,200 psi.

Subsequently, Kirchner, Gruver, Platts, and Walker (1967) and
Kirchner, Gruver, Platts, Rishel, and Walker (1968) attempted to
strengthen both conventionally sintered and reactively hot pressed
magnesia.  Silane treatments, decomposition of silicone grease and
packing in silicon carbide were used.  Small increases in strength
were observed in a few cases but the strengths did not approach
those achieved by Rhodes et al.

### 3.4 LOW EXPANSION COATINGS

In this section, processes for strengthening by low expansion
coatings are discussed.  These treatments are distinguished from
those in the previous section because the coating phase or phases
do not form by reaction with the specimen.  Chemical reaction with
the specimen is only necessary for adherence.

Moody (1969) used $B_4C$ coatings to strengthen titanium carbide
bodies.  $B_4C$ has a lower thermal expansion coefficient than titanium

carbide ($56 \times 10^{-7}$°C$^{-1}$ vs $80 \times 10^{-7}$°C$^{-1}$). $B_4C$ powder was applied
to the surface and sintered. A 50% increase in strength to 10,300
psi was observed. The titanium carbide specimens were very weak
to begin with so that even after strengthening the material was
not as strong as other available materials.

Kirchner, Buessem, Gruver, Platts, and Walker (1970) and
Kirchner, Gruver, and Platts (1971) attempted to strengthen the
following materials using coatings formed by chemical vapor deposition:

| Body | Coating |
|---|---|
| $ZrB_2$ + SiC (hot pressed) | CVD - SiC |
| KT-SiC* | CVD - SiC |
| SiC (hot pressed) | CVD - SiC |
|  | CVD - $Si_3N_4$ |
| Zircon porcelain | CVD - SiC |
|  | CVD - $Si_3N_4$ |

In a few cases, small improvements in strength were observed.
KT-SiC was about 30% stronger with a CVD SiC coating. In this
case there is no reason to expect compressive surface stresses to
be induced so the strengthening effect may be caused by the
strength of the coating or healing of surface flaws. Zircon
porcelain was strengthened by up to 40% (to 41,400 psi) using
CVD-$Si_3N_4$ coatings. The strength increased with increasing coating
thickness up to at least 11 μm. In this case compressive surface
stresses were expected because the thermal expansion coefficient
of $Si_3N_4$ is lower than that of zircon porcelain ($26 - 31 \times 10^{-7}$°C$^{-1}$
vs $41 \times 10^{-7}$°C$^{-1}$). Hot pressed SiC was strengthened slightly
by nitriding CVD-Si coatings to form $Si_3N_4$, in one case by 18% to
49,830 psi. In this case, compressive surface stresses were
expected based on the thermal expansion differences.

---

*KT-SiC Carborundum Co., Niagara Falls, New York.

## 3.5  PHASE TRANSFORMATIONS

The use of phase transformations to induce compressive surface
stresses is promising because of the relatively large volume changes
that occur in some cases and because the stresses will be retained
in any temperature range in which the phase or phases remain stable.
Despite the advantages of this method very little work has been
done to develop these processes for strengthening polycrystalline
ceramics.

Kirchner, Gruver, Platts, Rishel, and Walker (1968) developed
a process using the tetragonal to monoclinic transformation to
strengthen zirconia.  In commercial zirconia bodies, additions
of CaO, MgO, or $Y_2O_3$ are used to partially stabilize the zirconia
in the cubic form.  The unstabilized portions of the body invert
from the tetragonal to the monoclinic form during cooling through
the transformation range ($\sim$1000°C).  There is a volume increase
of about 9% during this phase transformation.  If the stabilization
of the surface is decreased, the volume increase due to the transformation increases and compressive surface stresses are induced.
Silica was used to destabilize the surface by competing with the
zirconia for the stabilizer.

Zirconia powder[*], containing CaO and MgO as stabilizers, was
dry pressed and sintered at 1650°C for one hour.  Bars, approximately 0.095 x 0.235 x 1.75 in, were cut from these samples.  The
bars were packed in silicon carbide abrasive grain and refired.
Rod tests were used to demonstrate the presence of residual compressive stresses in the treated specimens (the thickness of the
slotted bars decreased by $\sim$0.011 in).  X-ray diffraction analysis
of the specimen surfaces showed that the surface material was
destabilized by the treatment.  Electron microprobe measurements
showed substantial penetration of $SiO_2$ to a depth of at least 0.007
in.

---

[*]Zirconia R, TAMCO, Niagara Falls, New York.

The flexural strengths of specimens treated at various combinations of time and temperature were measured.  The best results are given in Table 3.16.  A 42% increase in strength was observed for specimens packed in 120-mesh SiC and refired at 1300°C for 8 hours. At lower refiring temperatures and shorter times the strengths were lower because of insufficient reaction.  At high refiring temperature (1500°C) the reaction layer was very thick and had a poor structure leading to low strength.

A somewhat similar process was recently developed by Pascoe and Garvie (1977).  They found that in zirconia with metastable tetragonal domains, these domains can be transformed to the stable monoclinic phase by abrasion of the surface.  Compressive stresses are generated in the surface by the volume expansion and significant strength increases are observed.

Moody (1969) attempted to strengthen sintered fused silica using the high-low cristobalite transformation.  Cristobalite has a 5% volume decrease during cooling at about 270°C.  He enhanced cristobalite formation in the interior of sintered fused silica bodies by varying the purity and particle size of the powders. The strength of his specimens did not increase significantly but the scatter of the results was substantially reduced in some cases.  Rod tests qualitatively demonstrated the presence of potentially useful prestress.

It is interesting that under typical conditions the concentration of cristobalite in sintered fused silica bodies is greater at the surface than in the interior because the surface is exposed to impurities and other influences.  Therefore, one would expect typical bodies to have tensile surface stresses and resulting strength degradation.  It is evident that care should be taken to reduce cristobalite formation in these surfaces.

Table 3.16. Flexural Strength of Zirconia with Surface Layers Destabilized by Packing in SiC

| Treatment | No. Specimens | Flexural Strength Data* Average Flexural Strength psi | Strength Difference psi |
|---|---|---|---|
| Controls, as cut | 4 | 20,600 | - - - |
| Controls, refired 1300°C, 8 hours | 3 | 23,200 | +2,600 |
| Packed in 600-mesh SiC, 1300°C, 8 hours | 4 | 26,200 | +5,600 |
| Packed in 120-mesh SiC, 1300°C, 8 hours | 2 | 29,300 | +8,700 |

*Three point loading on a one inch span.

### 3.6  Status of Research on Coatings, Chemical Treatments
### and Phase Transformations

As in the case of strengthening by thermal treatments, there
are several questions commonly asked about strengthening by coatings, chemical treatments, and phase transformations.  Among these
questions are the following:

1.  Won't the internal flaws become fracture origins under the
    combined influence of the residual tensile stresses and
    stresses due to the applied loads, preventing useful increases
    in load carrying ability?
2.  Even if it is possible to obtain residual compressive surface
    stresses after cooling, by the use of low expansion surface
    layers, won't the stresses be relieved when the specimens
    are reheated?
3.  Even it it is possible to obtain residual compressive surface
    stresses by chemical treatments, won't the stresses be relieved
    by continued diffusion to reduce the composition gradient,
    or by other chemical instabilities?

Answers to these questions are given in the following paragraphs.
The answer to the first of these questions is essentially
the same as that given to a similar question regarding thermal
treatments in Section 2.6.  One difference is that the compressive
layers in chemically treated specimens are usually much thinner
than those resulting from quenching so that, for a given maximum
compressive stress, the tensile stresses in the interior are lower.
Therefore, chemically treated specimens should be less likely than
quenched specimens to have limited strengthening because of shifting of fracture origins from surface to internal flaws.  Secondly,
because the surface layers are usually thin, the maximum tensile
stress just under the surface layer is only slightly less than the
nominal stress at the surface.  Therefore, in this case improvements in flexural strength are a good indication that the strength
has been improved rather than that the load  carrying ability has
been increased simply by shifting of the stress profile.

It is certainly true that residual compressive stresses
induced in low expansion surface layers during cooling are
relieved when specimens are reheated.  This question brings out
the principle that the treatment must be chosen in relation to
the conditions required in the particular application.  The maximum
residual stresses should, if possible, exist under the conditions
in which the maximum external stresses are applied.  If those
stresses are applied at high temperatures, the treatment should
be designed to yield high residual stresses at high temperatures,
perhaps by applying high expansion surface layers at a low or
intermediate temperature.

The answer to the question about relief of residual stresses
by diffusion is that it may happen in some cases of prolonged
heating at high temperatures.  There were some possible indications
of this effect when specimens were treated at high temperatures and
subsequently were found to be weaker than other specimens treated
at somewhat lower temperatures.  On the other hand, diffusion
rates in most refractory ceramics are low.  In cases such as
$Al_2O_3$-$Cr_2O_3$ solid solution surface layers on alumina, it is a
substantial problem to obtain layers that are thick enough without
using treatment temperatures that are so high as to cause unaccept-
able grain growth.  Because the treatment temperatures are
typically several hundred degrees celsius above expected use
temperatures, diffusion rates are expected to be low and strengths
should be retained for long times.

### 3.6.1   Summary of Successes and Failures

The treatments listed in Table 3.17 are those considered to
be successful based on available information.  A certain degree
of subjective judgment was involved in some of the choices because
of the limited amount of available information.  The uncertainties
are of two types  (1) the reliability of the observed differences
in strength and  (2) whether the observed strengthening resulted
from the compressive surface stress mechanism.

Table 3.17.  Successful Treatments Using Coatings, Chemical Treatments, and Phase Transformations

| Type of Treatment | Treatment | Body Treated |
|---|---|---|
| Low expansion glaze | glaze | alumina |
| Ion exchange glazing | ion exchange using $Na^+$ or $K^+$ | alumina |
| Low expansion solid solution surface layers | $Al_2O_3$-$Cr_2O_3$ | alumina |
| | $Al_2O_3$-$Cr_2O_3$ | sapphire |
| | $TiO_2$-$SnO_2$ | titania |
| | $MgAl_2O_4$-$MgCr_2O_4$ | spinel |
| | $MgO$-$NiO$ | magnesia |
| | $MgO$-$CoO$ | magnesia |
| | $MgO$-$Fe_2O_3$ | magnesia |
| | $MgO$-$NiO$ | periclase |
| | $MgO$-$MnO$ | periclase |
| | $MgO$-$NiO$ | nickel oxide |
| | $Si$-$Al$-$O$-$N$ | silicon nitride |
| Compound surface layers formed by reaction with body | calcium aluminates | alumina |
| | mullite | alumina |
| | Li-Al-spinel | alumina |
| | $CoAl_2O_4$ | alumina |
| | calcium aluminates | sapphire |
| | mullite | sapphire |
| | forsterite | magnesia |
| Low expansion coatings | $B_4C$ | titanium carbide |
| | $Si_3N_4$ | zircon porcelain |
| | $Si_3N_4$ | silicon carbide |
| Phase transformation | less stabilized $ZrO_2$ | zirconia |

Perhaps half of the treatments attempted have been failures so far.  However, continued development can be expected to lead to significant strengthening in many of these cases.  In some cases the material combinations are less favorable than in those considered to be successes.  Among the factors involved are  (1) insufficient thermal expansion difference between the surface layer and the body,  (2) treatment temperatures that are too high leading to degradation of the body,  (3) non-uniform treatments, etc.  There was a tendency to emphasize work on systems that initially yielded good results so that many of the less favorable combinations received little attention.

### 3.6.2  Summary of the Advantages of Compressive Surface Stresses Induced by Coatings, Chemical Treatments and Phase Transformations

Thermal treatments usually yielded greater strength increases than those observed for coatings, chemical treatments, and phase transformations.  Does it make sense to consider these treatments for further development or for practical applications?  The answer is yes for several reasons.  One important reason is that if several strengthening processes yielding adequate strength are available, the ultimate choice can be made to take advantage of secondary characteristics such as hardness, corrosion resistance, smoothness, color, luster, etc.  Another reason is that quenched specimens may be best used in flexure at low and moderate temperature but may be vulnerable to relaxation of stresses at high temperatures or to failure in tension due to high internal tensile residual stresses.  Other treatments may avoid these difficulties.

Each type of treatment; coatings, chemical treatments and phase transformations, has particular advantages and disadvantages.  Coatings, including glazes, tend to have a sharp transition between the surface layer and the main body.  Shear failures may occur at these transitions.  On the other hand, coatings provide a greater variety of surface properties than can be obtained using the other

treatments.  These properties include the smoothness and luster
of glazes and the hardness and inertness of CVD carbides and
nitrides.

Chemical treatments include solid solution surface layers and
compound surface layers formed by reaction with the body.  The
solid solution surface layers have the advantage that there is no
sharp transition so that shear failures are unlikely between the
surface layer and the body.  In some cases, diffusion processes
lead to pore formation which may limit the strength improvement.
Another disadvantage is that stresses may be relieved by diffusion
or reevaporation which may occur at high temperatures as in the
cases of $Al_2O_3$-$Cr_2O_3$ surface layers on alumina or MgO-NiO surface
layers on magnesia.  Compound surface layers have many of these
same disadvantages.

Phase transformations have a very great potential advantage
in that the induced stresses should be stable so long as the phases
are stable.  In other words the stresses will not vary with temperature as they do for the treatments based on thermal expansion
differences.  More work is needed to enhance these improvements
and to demonstrate this advantage.

### 3.6.3   Recommendations for Research

The recommendations for research on quenching, presented in
Section 2.6.4 apply equally well to coatings, chemical treatments
and phase transformations and will not be repeated here.  In
addition to those recommendations, there is need for basic research
specifically related to this group of treatments.  One example is
the need for thermal expansion data for solid solutions.  Although
the author has made many such measurements, the number of measurements still to be made is very great.  In order to have a better
understanding of the mechanism of strengthening, it is highly
desirable to have an intensive investigation of the fracture of
specimens with one particular type of treatment.  If the variation
of fracture stresses and characteristics of fracture origins were

known for specimens of varying surface stresses and surface layer thicknesses subjected to various stress states, further process development could be planned more effectively.

Process development should continue to improve the properties of materials previously strengthened and to adapt these techniques to strengthening new materials. The method using phase transformations should receive special emphasis because of the possibility that the beneficial stresses can be retained over a wide range of temperatures.

# CHAPTER 4

## POTENTIAL APPLICATIONS

### 4.1 APPLICATIONS

Except for compressive glazes, the strengthening processes
described in the foregoing sections have not yet been used in
practical applications as far as we know.  Therefore, practical
applications of these processes can best be thought of in terms
of analogies with glass and glass-ceramic articles with compressive
surface stresses which have developed viable markets.  Among these
applications are dinnerware, drinking glasses, eye glasses, and
windows of camping trailers.

Applications for ceramics with compressive surface stresses
should be sought in the following areas:

| | |
|---|---|
| bearings | extrusion dies |
| cutting tools | pump parts |
| gas turbines | lighting envelopes |
| radomes | laser windows |
| irdomes | leading edges |
| armor | |
| other wear resistance applications | |

### 4.2 LIMITATIONS OF TREATMENTS

In general, the quenching treatments can be thought of as
size and shape limited.  If the particular component is made of
a ceramic such as alumina which can be effectively strengthened
by quenching and if the size and shape are amenable to quenching,
it is likely that the highest strengths can be achieved by this

process.  If the particular component is not suitable for quenching
for one or more of the reasons given above, coatings, chemical
treatments or phase transformation treatments should be considered.  Glazes can be applied to bodies of all shapes and sizes but
chemical vapor deposited surface layers may have limitations with
respect to uniformity on large components and internal surfaces.
The chemical treatments applied by packing and refiring, yield very
uniform treatments with few size or shape limitations evident at
present.  Therefore, these treatments should be given first prior-
ity in difficult cases.

## 4.3  DESIGN CONSIDERATIONS

Design with brittle materials involves many complexities.  It
is seldom adequate to simply apply a safety factor to the results
of strength measurements in order to derive an allowable design
stress limit for a particular material.  Three important factors
affecting the stresses that can be sustained by a component are
(1) the strength degradation expected because of subcritical crack
growth,  (2) uncertainty in the strength of particular specimens
because of scatter in the measured strengths and  (3) the possibility of strength degradation as a result of surface damage, grain
growth or other phenomena during processing or use.  The role of
compressive surface stresses in relation to these factors is discussed in the following sections.

### 4.3.1  Subcritical Flaw Growth

The rate of flaw growth (V) in ceramics is frequently repre-
sented by

$$V = A \ K_I^n \qquad\qquad\qquad (4.1)$$

in which A is a constant, $K_I$ is the stress intensity factor and
n represents the slope of plots of log V vs log $K_I$.  Values of n

range from 10-100 for various ceramics indicating that the crack
growth rate is strongly dependent on $K_I$. This strong dependence
can be viewed as an advantage because small reductions of $K_I$ below
the values at which crack growth rates are too high lead to substantial reductions in V.

For a given flaw of initial depth $(a_o)$, the rate of flaw
growth can be integrated over time to yield the flaw depth (a)
which in turn can be used to predict the fracture stress using

$$\sigma_F = \frac{Z}{Y} \frac{K_{IC}}{a^{1/2}} \tag{4.2}$$

Compressive surface stresses are an effective means for
reducing tensile stresses at surface flaws. Kirchner and Walker
(1971) demonstrated the advantages of this method for alumina
ceramics for which the times to failure, tested in flexure at
room temperature, were increased by an estimated factor of $10^{15}$.
Similar improvements should be expected in other cases.

### 4.3.2  Scatter in the Strengths

The scatter in the strengths of ceramics has been the subject
of many investigations and several methods have been proposed to
deal with this problem. These methods have included improvement
of processing to reduce the scatter, use of statistical methods
to predict the probabilities of failure of components of various
sizes with various stress distributions, and use of proof testing
to remove the weak components from the distribution. Wiederhorn
(1974b) and Wiederhorn, Fuller, Mandel, and Evans (1975) have
provided up-to-date treatments of the combined use of statistical
methods and proof testing to improve the reliability of ceramic
components.

Compressive surface stresses can be helpful in several ways.
In some cases the surface treatments are effective in reducing
scatter. An outstanding example was given for alumina in Figure
2.5. Achieving distributions of this kind, reliably, would make

it unnecessary to use statistical methods to predict the strength
distributions of groups of components.

Compressive surface stresses also decrease the tendency to
fail at surface flaws and increase the tendency to fail at internal
flaws.  If the tendency to fail at internal flaws can be shifted
to the point at which almost 100% of the failures originate at
internal flaws, it may not be necessary to account for surface
flaw failure.  In that case, analysis of the distribution of the
fracture stresses of a group of components should be much simpler.

In proof testing, the suppression of surface flaw failures
by compressive surface stresses allows the internal flaws to be
subjected to higher stresses without failure during proof testing.
At the very least this should improve the selection rate at the
proof testing step.  Also, it is probable that in many cases the
proof test stress can be so much higher than the stresses expected
in service that the probability of failure in service becomes
vanishingly small.

### 4.3.3  Surface Damage

As shown by Gruver and Kirchner (1973), compressive surface
stresses are effective in reducing penetration of surface damage
in polycrystalline ceramics.  These particular experiments involved
scratching alumina with a diamond point.  The remaining flexural
strengths of scratched specimens with compressive surface stresses
were much greater than those of specimens without compressive
surface stresses.  Kirchner and Miller (1974) indented alumina
ceramics with compressive surface stresses induced by quenching
using spherical indenters to form Hertz cracks.  The loads required
to degrade the remaining strengths of the alumina specimens with
compressive surface stresses below the strength of the as-received
alumina were twice the loads required to degrade the strengths of
the as-received material showing clearly the advantage of compressive surface stresses.

Marshall, Lawn, Kirchner, and Gruver (1978) used Vickers' indenters to damage 96% alumina rods strengthened by quenching. The remaining strengths of the treated specimens were substantially greater than the remaining strengths of untreated specimens subjected to similar loads.

In view of the fact that the Hertz theory applies to impact at low velocities, it is reasonable to expect a similar advantage for resistance to localized impact damage of specimens with compressive surface stresses. As mentioned previously, Gebauer, Krohn, and Hasselman (1972) have analyzed the advantages of compressive surface stresses in improving the thermal shock resistance and reducing the resulting strength degradation of ceramics. Thus, it is clear that compressive surface stresses can be helpful in reducing the strength degradation resulting from various types of surface damage.

### 4.3.4  Efficiencies of Energy Conversion Equipment

The efficiencies of energy conversion equipment are strongly dependent on the peak temperatures and the energy losses, such as those due to cooling air used to prevent overheating of vulnerable parts. Therefore, there is considerable interest in substituting ceramics for refractory metals in equipment such as gas turbines (van Reuth, 1974). Raising the inlet gas temperature of a gas turbine raises the Carnot efficiency. Use of ceramics may also permit reduction or elimination of the use of cooling gas. Therefore, if other factors remain approximately the same the efficiency of gas turbines should be increased substantially by the substitution of ceramics for metals. It may be possible, in many cases, to enhance the survival of ceramics used under these difficult conditions by using compressive surface stresses.

### 4.3.5  Weight Reduction

The specific gravities of refractory metals range from about
7 to 16.  Obviously, if ceramics with specific gravities, usually
ranging from 2.5 to 4.0, can be substituted for refractory metals,
an opportunity for substantial weight reduction exists.  The advantages of weight reduction can be particularly important in rotating
parts such as rotors in gas turbines in which the stresses depend
partly on centrifugal forces.  It will be doubly important in
cases in which subcritical crack growth or creep limits the life
of the component.

As shown in the previous sections, the strengths of the
strengthened ceramics are often comparable to those of structural
metals.  Because of lower densities the strength to weight ratios
are favorable.  The same is true of stiffness to weight ratios.
Weight reduction can be achieved because of these favorable ratios.
The advantages of these weight reductions may be greatest in cases
such as transportation equipment for which the lifetime costs
include large expenditures for fuel which are strongly related to
the weight of the vehicle and the fuel that must be carried to
move that weight.

### 4.4  COSTS

The cost of treatments to form compressive surface stresses
can be thought of mainly in terms of adding one more step to the
several steps normally required to produce a finished ceramic
article.  In general, the treatments have characteristics that
are similar to the other steps so that costs should be similar
to other individual processing steps and should be expected to be
only a small percentage of the total cost of producing the article.
We do not know of any less expensive way to achieve similar
improvements in strength.

As in all materials processing, the selection rate (percentage
of parts passing inspection) is of crucial importance.  Even a
small percentage lost at each step can add up to substantial losses.

In a 5 step process a selection rate of 90% at each step yields
only 59% good parts; a 10 step process only 35% good parts.  Because
the treatments to form compressive surface stresses are done near
the end of the overall processing and the parts have accumulated
the costs of earlier processing steps and process losses, it is
especially important to achieve a high selection rate for these
treatments.

## Costs of quenching processes

The cost of quenching processes for metals has been treated
briefly (Anonymous, 1964).  Although the temperatures required for
quenching ceramics are much higher than for metals, the article
at least points out some factors that are important in commercial
operations.  The cost of the quenching medium may be important
especially if expensive media such as silicone oil are used.  Use
of emulsions rather than pure oils has several advantages as
pointed out in Section 2.1.1.  The water used in the emulsion
increases the volume of the quenching medium sometimes by factors
of 10 or 20.  In addition, use of emulsions may reduce discolora-
tion of the surface and decomposition of the oil.

Quenching systems, through which substantial production passes,
require cooling to control the quenching medium temperature.  At
least two methods can be used, a cooling tower or a water cooled
heat exchanger.  Here again emulsions have an advantage because
boiling of the water limits the temperature rise and in some
circumstances, use of emulsions may avoid the need for a cooling
system.

The cost of heating the parts prior to quenching will depend
strongly on the method of heating used, the number of parts
heated in each batch, and the quenching temperature.  Electric
resistance heating can be used for temperatures up to about 1500°C.
For higher temperatures induction heating or gas-oxygen furnaces
are likely to be used.  It should be feasible with any of these
methods to heat several parts in a batch and then quench them
separately.

## Costs of packing treatments

The packing treatments involve packing the ceramic articles
in refractory powders which in turn are contained in refractory
boxes, and heating these assemblies for a controlled cycle.  Thus,
the equipment requirements are mainly a furnace capable of reaching
the highest temperatures needed, refractory boxes, and powders.
Usually the powders and boxes should be reusuable but some components such as decomposable fluoride powders would have to be
replaced each time.

The packing treatments are inherently economical because large
numbers of parts can be treated in each firing.  Assuming that
substantial numbers of parts are to be treated and that the powders can be reused, it should be possible to reduce the treatment
costs to less than $1.00 per part in most cases.

## Cost justification

It is evident that the strengthening treatments will add
incrementally to the cost of producing ceramic components and
that this cost increase must be justified by other life cycle cost
reductions or by improved performances.  In the case of substitution
of ceramics for refractory metals, the life cycle costs may be
reduced because of lower costs of the ceramics, greater efficiency
because of higher operating temperatures or reduced weight, etc.
In the case of improved ceramic cutting tools, the added cost might
be justified by an increase in the number of pieces machined by
each cutting tool and the  decreased cost of changing tools.  Alter-
natively, it may be possible to increase the cutting speed, thus
saving labor and machine time.  In other cases it may be possible
to substitute a less expensive strengthened ceramic for a more
expensive untreated ceramic.  For example, there might be an
application for a ceramic to be loaded in flexure to 75,000 psi
at room temperature.  If hot pressed alumina were  presently used,
it might be possible to substitute quenched 96% alumina for the
hot pressed alumina and save money.

# REFERENCES

Anonymous (1964), "Quenching and Martempering," American Society for Metals, Metals Park, Ohio, pages 191-194.

Badaliance, R., Krohn, D. A., and Hasselman, D. P. H. (1974), "Effect of Slow Crack Growth on the Thermal Stress Resistance of an NaO-CaO-SiO$_2$ Glass," J. Amer. Ceram. Soc. 57 (10) 432-436.

Barnett, R. L., Costello, J. L., Hermann, P. C., and Hofer, K. E. Jr. (1965), "The Behavior and Design of Brittle Structures," Illinois Institute of Technology Research Institute, Technical Report AFFDL-TR-65-165, Contract AF33 (615)-1494 (September, 1965).

Bickelhaupt, R. E. (1970), "Diffusional Prestressing of Ceramics," Southern Research Institute Summary Report A-415-2379-XI, Contract N00019-70-C-0159 (December, 1970).

Bickelhaupt, R. E. (1968), "Diffusional Prestressing of Ceramics," Southern Research Institute Summary Report 9222-2000-XII, Contract N00019-67-C-0518 (August, 1968).

Bickelhaupt, R. E., Cox, H. P., and Hurst, R. D. (1967), "Diffusional Prestressing of Ceramics," Southern Research Institute Summary Report 8504-1780-XX, Contract NO W 66-0132d (July, 1967).

Bortz, S. A. and Wade, T. B. (1967), "Analysis of Mechanical Test Procedure for Brittle Materials," Illinois Institute of Technology Research Institute Interim Report, AMRA CR C7-09/1.

223

Brown, W. F. Jr. and Srawley, J. E. (1966), "Plane Strain Crack
    Toughness Testing of High Strength Metallic Materials," ASTM
    Special Publication No. 410, American Society for Testing
    and Materials, Philadelphia.

Brubaker, B. D. and Russell, R. (1967), "Residual Stress Develop-
    ment in Laminar Ceramics," Bull. Amer. Ceram. Soc. 46 (12),
    1194-1197.

Buessem, W. R. and Gruver, R. M. (1972), "Computation of Residual
    Stresses in Quenched $Al_2O_3$," J. Amer. Ceram. Soc. 55 (2),
    101-104.

Burke, J. J., Reed, N. L., and Weiss, V., Editors (1966), "Strength-
    ening Mechanisms: Metals and Ceramics," Syracuse University
    Press.

Charles, R. J. and Shaw, R. R. (1962), "Delayed Fracture of
    Polycrystalline and Single Crystal Alumina," Gen. Elec. Res.
    Lab. Rept. 62-RL-3081 M.

Congleton, J., Petch, N. J., and Shiels, S. A. (1969), "Brittle
    Fracture of Alumina below 1000°C," Phil. Mag. 19 (160),
    795-807.

Conway, J. G. (1971), "Factors Influencing the Use of Ceramics in
    Deep-Submergence Applications," from Ceramics in Severe
    Environments, Edited by W. W. Kriegel and H. Palmour III,
    Plenum Press, New York, pages 477-488.

Coppola, J. A., Krohn, D. A., and Hasselman, D. P. H. (1972),
    "Strength Loss of Brittle Ceramics Subjected to Severe Thermal
    Shock," J. Amer. Ceram. Soc. 55 (9), 481.

Cottrell, A. H. (1964), "The Mechanical Properties of Matter,"
    John Wiley and Sons, New York.

Crouch, A. G. and Jolliffe, K. H. (1970), "The Effect of Stress
    Rate on the Rupture Strength of Alumina and Mullite Refrac-
    tories," Proc. Brit. Ceram. Soc. 15, 37-46.

Cutler, I. B. and Brown, S. D. (1964), "Effects of Impurities on
    the Properties of Ceramics," from Parikh, N. M., "Studies
    of the Brittle Behavior of Ceramic Materials," Illinois
    Institute of Technology Research Institute Technical Report
    ASD-TR-61-628, Part III, Contract AF33 (657)-10697, (June,
    1964).

Davidge, R. W. and Tappin, G. (1970), "The Effects of Temperature
    and Environment on the Strength of Two Polycrystalline
    Aluminas," Proc. Brit. Ceram. Soc. 15, 47-60.

Dawihl, Von W. and Klinger, E. (1966), "Sintered Alumina and the
    Dependence of its Aging on Surface Active Substances," Ber.
    Deut. Keram. Ges. 43, 473-476.

Doherty, J. E., Tschinkel, J. G., and Copley, S. M. (1973), "Improve-
    ment in Thermal Shock Resistance by Surface Prestressing,"
    Bull. Amer. Ceram. Soc. 52 (9) 681-686.

Duga, J. J. (1969), "Surface Energy of Ceramic Materials," Defense
    Ceramic Information Center Report, 69-2.

Dunsmore, G. K., Fenstermacher, J. E., and Hummel, F. A. (1961),
    "High Temperature Mechanical Properties of Ceramics Materials:
    III, Vitrified Chemical and Refractory Porcelains," Bull.
    Amer. Ceram. Soc. 40 (5), 310-313.

Evans, A. G. (1974), "Analysis of Strength Degradation After Sus-
    tained Loading," J. Amer. Ceram. Soc. 57 (9), 410-411.

Evans. A. G. (1974), "Fracture Mechanics Determinations," from
    Fracture Mechanics of Ceramics, Vol. 1, Edited by R. C.
    Bradt, D. P. H. Hasselman and F. F. Lange, Plenum Press,
    New York, pages 17-48.

Evans, A. G. and Tappin, G., (1972), "Effects of Microstructure
    on the Stress to Propagate Inherent Flaws," Proc. Brit. Ceram.
    Soc. 20, 275-97.

Frazier, J. T., Jones, J. T., Raghavan, K. S., McGee, T. D., and
    Bell, H. (1971), "Chemical Strengthening of $Al_2O_3$," Bull.
    Amer. Ceram. Soc. 50 (6), 541-544.

Gebauer, J. and Hasselman, D. P. H. (1971), "Elastic-Plastic
    Phenomena in the Strength Behavior of an Aluminosilicate
    Ceramic Subjected to Thermal Shock," J. Amer. Ceram. Soc.
    54 (9), 468-469.

Gebauer, J., Krohn, D. A., and Hasselman, D. P. H. (1972), "Thermal-
    Stress Fracture of a Thermomechanically Strengthened
    Aluminosilicate Ceramic," J. Amer. Ceram. Soc. 55 (4),
    198-201.

Geller, R. F., and Insley, H. (1932), "Thermal Expansion of Some
    Silicates of Elements in Group II of the Periodic System,"
    Bur. Stds. J. Res. 9, 35-46, RP-456.

Gielisse, P. J. and Stanislaw, J. (1970), "Dynamic and Thermal
    Aspects of Ceramic Processing," Univ. of Rhode Island Final
    Tech. Report, Contract N00019-70-C-0163.

Glenny, E. and Royston, M. G. (1958), "Transient Thermal Stresses
    Promoted by Rapid Heating and Cooling of Brittle Circular
    Cylinders." Trans. Brit. Ceram. Soc. 57 (10), 645-77.

Griffith, A. A. (1920), "The Phenomena of Rupture and Flow in
    Solids," Phil. Trans. Roy. Soc. 221A (4), 163-98.

Grosskreutz, J. C. (1970), "The Effect of Environment on the Frac-
    ture of Adhering Aluminum Oxide," J. Electrochem. Soc. (Solid
    State Science) 117, 94-943.

Grosskreutz, J. C. (1969), "Mechanical Properties of Metal Oxide
    Films," J. Electrochem Soc. (Solid State Science) 116, 1232-
    1237.

Gruszka, R. F., Mistler, R. E., and Runk, R. B. (1970), "Effect of
    Various Surface Treatments on the Bend Strength of High Alumina
    Substrates," Bull. Amer. Ceram. Soc. 49 (6), 575-579.

Gruver, R. M. and Buessem, W. R. (1971), "Residual Stresses in
    Cylindrical Rods as Measured by Dimensional Changes After
    Slotting," Bull. Amer. Ceram. Soc. 50 (9), 749-751

Gruver, R. M. and Buessem, W. R. (1969), "A Rod Test for Measuring
    Stresses in Ceramic Bodies with Compressive Surface Layers,"
    Ceramic Finishing Company Special Report, Contract N00019-69-
    C-0225.

Gruver, R. M. and Kirchner, H. P. (1973), "Effect of Surface Damage
    on the Strength of $Al_2O_3$ Ceramics with Compressive Surface
    Stresses," J. Amer. Ceram. Soc. 56 (1), 21-24.

Gruver, R. M. and Kirchner, H. P. (1968), "Residual Stress and
    Flexural Strength of Thermally Conditioned 96% Alumina Rods,"
    J. Amer. Ceram. Soc. 51 (4) 232.

Gruver, R. M., Platts, D. R., and Kirchner, H. P., "Strengthening
    Silicon Carbide by Quenching," Amer. Ceram. Soc. Bull. 53 (7),
    524-527.

Gruver, R. M., Sotter, W. A., and Kirchner, H. P. (1976), "The
    Variation of Fracture Stress with Flaw Character in 96% $Al_2O_3$,"
    Bull. Amer. Ceram. Soc. 55 (2) 198-202.

Guard, R. W. and Romo, P. C. (1965), "X-ray Microbeam Studies of
    Fracture Surfaces in Alumina," J. Amer. Ceram. Soc. 48(1),
    7-11.

Gupta, T. K. (1972), "Strength Degradation and Crack Propagation
    in Thermally Shocked $Al_2O_3$," J. Amer. Ceram. Soc. 55 (5),
    249-253.

Harrison, W. B. (1965), "Fabrication and Fracture of Polycrystalline NiO," Honeywell, Inc., Third Interim Technical Report, Contract DA-11-022-ORD-3441, (March, 1965).

Harrison, W. B. (1963), "The Influence of Surface Condition on the Strength of Polycrystalline MgO," Honeywell, Inc., First Interim Technical Report, Contract DA-11-022-ORD 3441 (April, 1963).

Hasselman, D. P. H. (1970), "Strength Behavior of Polycrystalline Alumina Subjected to Thermal Shock," J. Amer. Ceram. Soc. 53 (9), 490-495.

Heuer, A. H. (1969), "Transgranular and Intergranular Fracture in Polycrystalline Alumina," J. Amer. Ceram. Soc. 52 (9), 510-11.

Hockey, B. J. (1971), "Plastic Deformation of Aluminum Oxide by Indentation and Abrasion," J. Amer. Ceram. Soc. 54 (5), 223-231.

Hockey, B. J., and Lawn, B. R. (1975), "Electron Microscopic Observations of Microcracking About Indentations in Aluminum Oxide and Silicon Carbide," National Bureau of Standards Report NBSIR 75-658 (January, 1975).

Hummel, F. A. and Lowery, H. E. (1951), "Quenching Vitrious Bodies Adds Strength," Ceram. Ind. 56 (6), 93-94.

Inglis, C. E. (1913), Trans. Instn. Nav. Archit. 55, 219.

Insley, R. H. and Barczak, V. J. (1964), "Thermal Conditioning of Polycrystalline Alumina Ceramics," J. Amer. Ceram. Soc. 47 (1), 1-4.

Jacobson, L. A. and Fehrenbacher, L. L. (1966), "Surface and Microstructural Influence on the Flexure Strength of Dense Polycrystalline MgO," Air Force Materials Laboratory Technical Report, TR 66-91.

Kelly, A. (1966), "Strong Solids," Clarendon Press, Oxford.

Kingery, W. D. and Pappis, J. (1956), "Note on Failure of Ceramic Materials at Elevated Temperatures Under Impact Loading," J. Amer. Ceram. Soc. 39 (2), 64-66.

Kirchner, H. P. (1978), "The Strain Intensity Criterion for Crack Branching in Ceramics," Engineering Fracture Mechanics 10, 283-288.

Kirchner, H. P. (1974), "Strengthening of Oxidation Resistant Materials for Gas Turbine Applications," Ceramic Finishing Company Report, NASA CR 134661, Contract NAS3-16788, (June, 1974).

Kirchner, H. P. (1969), "Thermal Expansion Anisotropy of Oxides and Oxide Solid Solutions," J. Amer. Ceram. Soc. 52 (7), 379-386.

Kirchner, H. P., Buessem, W. R., Gruver, R. M., Platts, D. R., and Walker, R. E. (1970), "Chemical Strengthening of Ceramic Materials," Ceramic Finishing Company Summary Report, Contract N00019-70-C-0418, (December, 1970).

Kirchner, H. P., and Gruver, R. M. (1974), "The Elevated Temperature Flexural Strength and Impact Resistance of Alumina Ceramics Strengthened by Quenching," Mater. Sci. Eng. 13, 63-69.

Kirchner, H. P. and Gruver, R. M. (1973), "Fracture Mirrors in Alumina Ceramics," Phil. Mag. 27 (6), 1433-1446.

Kirchner, H. P. and Gruver, R. M. (1970), "Strength-Anisotropy-Grain Size Relations in Ceramic Oxides," J. Amer. Ceram. Soc. 53 (5), 232-236.

Kirchner, H. P. and Gruver, R. M. (1966), "Chemical Strengthening of Polycrystalline Ceramics," J. Amer. Ceram. Soc. 49 (6), 330-333.

Kirchner, H. P. and Gruver, R. M. (1965), "Chemical Strengthening of Ceramic Materials," Linden Laboratories Summary Report, Contract NOW-0381-C, (May, 1965).

Kirchner, H. P., Gruver, R. M., and Platts, D. R. (1971), "Chemical Strengthening of Ceramic Materials," Ceramic Finishing Company Summary Report, Contract N00019-71-C-0208 (December, 1971).

Kirchner, H. P., Gruver, R. M., Platts, D. R., Rishel, P. A., and Walker, R. E. (1969), "Chemical Strengthening of Ceramic Materials," Ceramic Finishing Company Summary Report, Contract N00019-68-C-0142 (January, 1969).

Kirchner, H. P., Gruver, R. M., Platts, D. R., Rishel, P. A., and Walker, R. E. (1968), "Chemical Strengthening of Ceramic Materials," Linden Laboratories Summary Report, Contract N00019-67-C-0489, (April, 1968).

Kirchner, H. P., Gruver, R. M., Platts, D. R., and Walker, R. E. (1967), "Chemical Strengthening of Ceramic Materials," Linden Laboratories Summary Report, Contract NO W 66-0441-C (April, 1967).

Kirchner, H. P., Gruver, R. M., and Sotter, W. A. (1976), "Characteristics of Flaws at Fracture Origins and Fracture Stress-Flaw Size Relations in Various Ceramics," Mater. Sci. Eng. 22, 147-156.

Kirchner, H. P., Gruver, R. M., and Sotter, W. A. (1975), "Use of Fracture Mirrors to Interpret Impact Fractures in Brittle Materials," J. Amer. Ceram. Soc. 58 (5-6), 188-191.

Kirchner, H. P., Gruver, R. M., and Sotter, W. A. (1974), "The Variation of Fracture Mirror Radius with Fracture Stress for Polycrystalline Ceramics under Various Loading Conditions," Ceramic Finishing Company, Technical Report No. 2, Contract N00014-74-C-0241, (November, 1974).

Kirchner, H. P., Gruver, R. M., and Walker, R. E., (1973), "Strengthening Hot Pressed $Al_2O_3$ by Quenching," J. Amer. Ceram. Soc. 56 (1), 17-21.

Kirchner, H. P., Gruver, R. M., and Walker, R. E., (1972), "Strength Effects Resulting from Simple Surface Treatments," from Science of Ceramic Machining and Surface Finishing, S. J. Schneider, Jr. and R. W. Rice, Editors, N.B.S. Special Publication 348, pp. 353-363.

Kirchner, H. P., Gruver, R. M., and Walker, R. E. (1969), "Strengthening Sapphire by Compressive Surface Layers," J. Appl. Phys. 40 (9), 3445-2452.

Kirchner, H. P., Gruver, R. M., and Walker, R. E. (1968), "Strengthening Alumina by Glazing and Quenching," Bull. Amer. Ceram. Soc. 47 (9) 798-802.

Kirchner, H. P., Gruver, R. M., and Walker, R. E. (1968), "Chemical Strengthening of Polycrystalline Alumina," J. Amer. Ceram. Soc. 51 (5), 251-255.

Kirchner, H. P., Gruver, R. M., and Walker, R. E., (1967), "Chemically Strengthened, Leached Alumina and Spinel," J. Amer. Ceram. Soc. 50 (4), 169-173.

Kirchner, H. P. and Miller, C. S., (1974), Unpublished research on the resistance of alumina ceramics to localized impact damage.

Kirchner, H. P. and Rishel, P. A. (1971), "Measuring the Tensile Strength of a Brittle Material Using a Thermal Contraction Loading Device," J. Mat. 6 (1), 39-47.

Kirchner, H. P., Scheetz, H. A., Brown, W. R., and Smyth, H. T. (1962), "Investigation of Theoretical and Practical Aspects of the Thermal Expansion of Ceramic Materials," Cornell Aeronautical Laboratory Final Report No. PI-1273-17-12, Contract NOrd-18419, (July, 1962).

Kirchner, H. P., Sotter, W. A., and Gruver, R. M. (1975),
    "Strengthening Hot Pressed $Si_3N_4$ by Heating and Quenching,"
    J. Amer. Ceram. Soc. 58 (7-8), 353.

Kirchner, H. P. and Walker, R. E. (1971), "Delayed Fracture of
    Alumina Ceramics with Comprssive Surface Stresses," Mater.
    Sci. Eng. 8, 301-309.

Kirchner, H. P., Walker, R. E., and Gruver, R. M., (1971),
    "Strengthening Alumina by Quenching in Various Media," J.
    Appl. Phys. 42 (10), 3685-3692.

Kistler, S. S. (1962), "Stresses in Glass Produced by Non-uniform
    Exchange of Monovalent Ions," J. Amer. Ceram. Soc. 45 (2),
    59-68.

Kohatsu, I. (1967), "Solid State Reactions between CaO and $Al_2O_3$,"
    M.S. Thesis, The Pennsylvania State University.

Lange, F. F. (1970), "Healing of Surface Cracks in SiC by Oxidation,"
    J. Amer. Ceram. Soc., 53 (5), 290.

Lange, F. F and Gupta, T. K. (1970), "Crack Healing by Heat Treat-
    ment," J. Amer. Ceram. Soc. 53 (1), 54-55.

Lange, F. F. and Radford, K. C. (1970), "Healing of Surface
    Cracks in Polycrystalline $Al_2O_3$," J. Amer. Ceram. Soc. 53
    (7), 420-421.

Lawn, B. R. and Wilshaw, T. R. (1975), "Fracture of Brittle Solids,"
    Cambridge University Press, London.

Leach, J. S. L. (1970), "The Plasticity of Thin Oxide Films,"
    Proc. Brit. Ceram. Soc. 15, 215-223.

Liu, T. S., Stokes, R. J., and Li, C. H. (1964), "Fabrication
    and Plastic Behavior of Single Crystal MgO-NiO and MgO-MnO
    Solid Solution Alloys," J. Amer. Ceram. Soc. 47 (6), 276-279.

Mallinder, F. P. and Proctor, B. A. (1966a), "The Strengths of
    Flame-Polished Sapphire Crystals," Phil. Mag. 13, 197-208.

Mallinder, F. P. and Proctor, B. A. (1966b), "Preparation of High
    Strength Sapphire Crystals," Proc. Brit. Ceram. Soc. (6),
    9-11.

Marshall, D. B. and Lawn, B. R. (1978), "Strength Degradation of
    Thermally Tempered Glass Plates," J. Amer. Ceram. Soc. 61
    (1-2), 21-27.

Marshall, D. B., Lawn, B. R., Kirchner, H. P., and Gruver, R. M. (1978), "Contact Induced Strength Degradation of Thermally Treated Al$_2$O$_3$," J. Amer. Ceram. Soc. 61 (5-6), 271-272.

Merz, K. M., Brown, W. R., and Kirchner, H. P. (1962), "Thermal Expansion Anisotropy of Oxide Solid Solutions," J. Amer. Ceram. Soc. 45 (11), 531-536.

Mohr, T. W. (1974), "A Better Way to Evaluate Quenchants," Metal Progress, 85-88.

Moody, W. E. (1969), "Prestressed Non-Oxide and Selected High Temperature Oxide Ceramic Structures," Georgia Institute of Technology Final Technical Report, Contract DA-AH01,67-C 1494, (June, 1969).

Morley, J. G. and Proctor, B. A. (1962), "Strengths of Sapphire Crystals," Nature, 196, 1082.

Mountvala, A. T. and Murray, G. T. (1964), "Effect of Gaseous Environment on the Fracture Behavior of Al$_2$O$_3$," J. Amer. Ceram. Soc. 47 (5), 237-239.

Nehring, V. W. and Jones, J. T. (1973), "Flexural Creep of Surface-Treated Sapphire Rods," J. Amer. Ceram. Soc. 56 (1), 50.

Neuber, H. (1958), "Theory of Notch Stresses: Principles for Exact Calculation of Strength with Reference to Structural Form and Material," Springer-Verlag, Berlin AEC Translation No. 4547.

Nordberg, M. E., Mochel, E. L., Garfinkel, H. M., and Olcott, J. S. (1964), "Strengthening by Ion Exchange," J. Amer. Ceram. Soc. 47 (5), 215-19.

Orr, L. (1972), "Practical Analysis of Fractures in Glass Windows," Materials Res. and Standards 12, 21-23 and 47.

Pascoe, R. T. and Garvie, R. C. (1977), "Surface Strengthening of Transformation-Toughened Zirconia," from Ceramic Microstructures, Edited by Richard M. Fulrath, Westview Press, Boulder, Colorado.

Pears, C. D. and Starrett, H. S. (1966), "An Experimental Study of the Weibull Volume Theory," Southern Research Institute Technical Rept. No. AFML-TR-66-228.

Pears, C. D., Starrett, H. S., Bickelhaupt, R. E., and Braswell, D. W. (1970), "A Quantitative Evaluation of Test Methods for Brittle Materials," Southern Research Institute Technical Report. AFML-TR-69,244, Part I and II.

Pearson, S. (1956), "Delayed Fracture in Sintered Alumina," Proc. Phys. Soc. (London), 69B, 1293-96.

Petrovic, J. J., Jacobson, L. A., Talty, P. K., and Vasudevan, A. K. (1975), "Controlled Surface Flaws in Hot-Pressed $Si_3N_4$," J. Amer. Ceram. Soc. 58 (3-4), 113-116.

Phillips, C. J. and DiVita, S. (1964), "Thermal Conditioning of Ceramic Material," Bull. Amer. Ceram. Soc. 43 (1), 6-8.

Platts, D. R. and Kirchner, H. P. (1971), "Comparing Tensile and Flexural Strengths of a Brittle Material," J. Mat. 6 (1), 48-59.

Platts, D. R., Kirchner, H. P., Gruver, R. M., and Walker, R. E. (1970), "Strengthening Glazed Alumina by Ion Exchange," J. Amer. Ceram. Soc. 53 (5), 281.

Rhodes, W. H., Berneberg, P. L., Cannon, R. M., and Stule, W. C. (1973), "Microstructure Studies of Polycrystalline Refractory Oxides," AVCO Corp. Summary Report, Contract N00019-72-C-0298, (April, 1973).

Rhodes, W. H., Sellers, D. J., Vasilos, T., Heuer, A. H., Duff, R., Burnett, P. (1966), "Microstructure Studies of Polycrystalline Refractory Oxides," Avco Corporation Summary Report, Contract NOW-65-0316-f (March, 1966).

Rice, R. W. (1974), "Fractographic Identification of Strength Controlling Flaws and Microstructure," from Fracture Mechanics of Ceramics, Vol. I, Edited by R. C. Bradt, D. P. H. Hasselman, and F. F. Lange, Plenum Press, New York, pages 323-345.

Rice, R. W. (1972),Unpublished research on fracture mirrors.

Rice, R. W. and McDonough, W. J. (1972), "Ambient Strength and Fracture of $ZrO_2$," Mechanical Behavior of Materials, Vol. IV, The Society of Material Science, Japan, Pages 394-403.

Rice,, R. W. and McDouough, W. J. (1972), "Ambient Strength and Fracture Behavior of $MgAl_2O_4$," Mechanical Behavior of Materials, Vol. IV, The Society of Materials Science, Japan, pages 422-431.

Rigby, G. R. and Green, A. T. (1943), "Thermal Expansion Characteristics of the Calcium Aluminates and Calcium Ferrites," Trans. Brit. Ceram. Soc. 42 (5), 95-103.

Rishel, P. A., Infield, J. M., and Kirchner, H. P. (1968), "Leaching and Machining of Polycrystalline Alumina," Bull. Amer. Ceram. Soc. 47 (8), 702-706.

Roszhart, T. V., Pearson, D. J., and Bohn, J. R. (1971a), "Holo-
    graphic Characterization of Ceramics," Part I, TRW Systems
    Group Report, Contract N00019-69-C 0228.

Roszhart, T. V. and Bohn, J. R. (1971b), "Holographic Character-
    ization of Ceramics (Observation of Static Fatigue)," TRW
    System Group Report, Contract N00019-70-C-0136.

Rudnick, A., Marshall, C. W., Duckworth, W. H., and Emerick, B. R.
    (1968), "The Evaluation and Interpretation of Mechanical
    Properties of Brittle Materials," Defense Ceramic Information
    Center Report 68-3, AFML-TR-67-361.

Ryshkewitch, E. (1960), "Oxide Ceramics," Academic Press, New York.

Sarkar, B. K. and Glenn, T. G. J. (1970), "Fatigue Behavior of
    High-Al$_2$O$_3$ Ceramics," Trans. Brit. Ceram. Soc. 69 (5), 199-203.

Schurecht, H. G. and Pole, G. R. (1930, "Method of Measuring
    Strains Between Glazes and Ceramic Bodies," J. Amer. Ceram.
    Soc. 13, 369-375.

Sedlacek, R. and Halden, F. A. (1962), "Method for Tensile Testing
    of Brittle Materials," Rev. Sci. Instr. 33 (3), 298-300.

Sedlacek, R. (1968), "Tensile Fatigue Strength of Brittle Materials,"
    Stanford Res. Inst. Rept. No. AFML-TR-66-245.

Semple, C. W. (1970), "Residual Stress Determinations in Alumina
    Bodies," Army Materials and Mechanics Research Center Report
    No. AMMRC TR 70-14, (June, 1970).

Skrovanek, S. D. and Bradt, R. C. (1976), "Surface Strengthening
    Treatment of a Reaction Bonded Silicon Nitride," Presented
    at the Annual Meeting, American Ceramic Society (May, 1976).

Smith, F. W., Emery, A. F., and Kobayashi, A. S. (1967), "Stress
    Intensity Factors for Semicircular Cracks, Part 2 - Semi-
    Infinite Solid," J. Appl. Mechanics 34, Series E, 953-959.

Smith, F. W., Kobayashi, A. S., and Emery, A. F. (1967), "Stress
    Intensity Factors for Penny-Shaped Cracks, Part 1 - Infinite
    Solid," J. Appl. Mechanics 34, Series E, 947-952.

Smoke, E. J. and Koenig, J. H. (1958), "Thermal Properties of
    Ceramics," Rutgers, The State University, Engineering
    Research Bulletin No. 40.

Specian, G. and Hasselman, D. P. H. (1975), "Bibliobraphy of the
    Thermal Stress Fracture of Ceramics, Glasses and Refractories,"
    Ceramics Res. Lab., Lehigh U. (August, 1975).

Spriggs, R. M., Brissette, L. A., and Vasillos, T. (1964), "Pressure Sintered Nickel Oxide," Bull. Amer. Ceram. Soc. 43 (8), 572.

Stookey, S. D. (1965), "Strengthening Glass and Glass-Ceramics by Built-In Surface Compression," from High Strength Materials, Edited by V. F. Zackay, John Wiley, New York.

Tada, H., Paris, P. C., and Irwin, G. R. (1973), "The Stress Analysis of Cracks Handbook," Del Research Corporation, Hellertown, Pennsylvania.

Tressler, R. E., Langensiepen, R. A., and Bradt, R. C. (1974), "Surface-Finish Effects on Strength-vs-Grain-Size Relations in Polycrystalline $Al_2O_3$," J. Amer. Ceram. Soc. 57 (5), 226-227.

Van Reuth, E. C. (1974), "The Advanced Research Project Agency's Gas Turbine Program," from Ceramics for High Performance Applications, Edited by J. J. Burke, A. E. Gorum and R. N. Katz, Brook Hill Publishing Co., Chestnut Hill. Mass.

Wachtman, J. B. Jr. and Maxwell, L. H. (1954), "Plastic Deformation of Ceramic Oxide Single Crystals," J. Amer. Ceram. Soc. 37 (7), 291-299.

Warshaw, S. I. (1957), "Prestressed Ceramics," Bull. Amer. Ceram. Soc. 36 (1), 28.

Weymann, H. D. (1962), "A Thermoviscoelastic Description of Tempering of Glass," J. Amer. Ceram. Soc. 45 (11), 517-522.

Wiederhorn, S. M. (1974a), "Subcritical Crack Growth in Ceramics," from Fracture Mechanics of Ceramics, Volume 2, Plenum Press, New York, pages 613-646.

Wiederhorn, S. M. (1974b), "Reliability, Life Prediction and Proof Testing of Ceramics," from Ceramics for High Performance Applications," J. J. Burke, A. E. Gorum, and R. N. Katz, Editors, Brook Hill Publishing Company, Chestnut Hill, Mass., pages 633-663.

Wiederhorn, S. M. (1969), "Fracture of Sapphire," J. Amer. Ceram. Soc. 52 (9), 485-491.

Wiederhorn, S. M., Fuller, E. R., Mandel, J., and Evans, A. G. (1975), "An Error Analysis of Failure Prediction Techniques," Paper 53-BE-75-F Fall Meeting, Basic Science Div. Amer. Ceram. Soc., (September, 1975), Abstract in Bull. Amer. Ceram. Soc. 54 (8), 742.

Wiederhorn, S. M., Hockey, B. J., and Roberts, D. E. (1973),
    "Effect of Temperature on the Fracture of Sapphire," Phil.
    Mag. 28 (4), 783-796.

Wiederhorn, S. M., Johnson, H., Diness, A. M., and Heuer, A. H.
    (1974), "Fracture of Glass in Vacuum," J. Amer. Ceram. Soc.
    57 (8), 336-341.

Williams, L. S. (1956), "Stress Endurance of Sintered Alumina,"
    Trans. Brit. Ceram. Soc. 55, 287-312.

# INDEX

## A

Abrasion, 42
  of alumina, 42, 73
  of sapphire, 115
Atomic lattice planes, 1
Alumina, 27, 31
  hot pressed, 81-110
    flexural strength of,
      82-89
    fracture mirrors, 98
    grain size of, 86
    quenching of, 81-110
    specimen preparation, 81
    strength degradation of,
      91-96
    stress profiles, 95-103
  96% alumina, 27
    $Al_2O_3$-$Cr_2O_3$ treatment of,
      163-173
    compound surface layers
      on, 188
    flexural strength of,
      27, 41, 49
    glazing and quenching
      of, 31-80
    quenching of, 31-80
    ring test, 38
    rod test, 38
    tensile strength of,
      27, 49,
  sintered, 31-80
    compound surface layers
      on, 188

Aluminosilicate ceramic (see
    mullite)
  quenching of, 51, 54
Applications, 215-222
  design considerations, 216
  efficiency of energy use, 219
  limitations of treatments, 215
  scatter in the strengths, 217
  subcritical flaw growth, 216
  surface damage, 218
  weight reduction, 220

## C

Ceramics, 1
Characterization, 20, 38
  techniques of, 20
  treated specimens, 38
Charpy impact test, 28, 60
Chemical polishing, 110, 115
Cleavage, 10
  reflecting spots, 104
Coatings, low expansion, 203
  of silicon carbide, 204
  of titanium carbide, 203
  of zircon porcelain, 204
  of zirconium deboride, 204
Composition profile, 112
  $Al_2O_3$-$Cr_2O_3$ solid solution,
      168, 171, 173
  calcium aluminate, 190
  glazed sapphire, 112

Composition profile [Cont.]
  ion exchange glazed alumina,
      163
  $MgCr_2O_4$-$MgAl_2O_4$ solid
      solution, 179
  mullite, 196
  sapphire, 199
  $TiO_2$-$SnO_2$ solid solution,
      177
  zirconia, 205
Compound surface layers, 14,
      188-203
  on alumina, 188
  on sapphire, 199
Compressive surface stress,
      12, 18, 29
Compressive surface force,
      20, 38, 51
Cost, 220
  of quenching processes, 221
  of packing treatments, 222
  justification, 222
Crack, 3
  branching, 23
  velocity, 9, 117
Critical stress intensity
      factor, 23


                        D

Delayed fracture, 13, 31,
      65-72
  of alumina, 65
    effect of humidity, 66
    stress corrosion
        mechanisms, 72
    surviving specimens, 68
Design considerations, 216
Drop weight impact test, 28


                        E

Efficiency of energy use, 219
Electrical porcelain, 128
  glazing and quenching of,
      128, 130
  quenching of, 128, 130
Emulsifiers, 34, 36

Environmental effects, 9
  delayed fracture, 31
  humidity, 66


                        F

Flame polishing, 1, 12, 110, 115
Flaw, 1, 102
  abrasion, 42
  characteristics, 6
  fracture origin, 16
  healing, 55, 69, 113, 119,
      134
  linking, 6
  location of, 1
    surface, 14, 104
    internal (volume), 1, 14
  shape parameter, 5
  types, 6, 102
    crystals, large, 6, 106
    elliptical, 2
    machining, 8
    penetration, 6, 9, 104
    pore, 6, 102
    porous regions, 8, 103
    stepped, 6, 9
Fluorine treatments, 45, 121
      (*see also* Leaching)
Forsterite, 120
  glazing of, 126
  glazing and quenching of,
      126
  surface layer, 203
Fracture, 4
  energy, 4
  mechanics, 4, 23, 117
  mirror, 6, 15, 22, 92, 95,
      97, 106, 109, 134, 141
  origins, techniques for
      locating, 16, 97
  stress, 24


                        G

Glaze, 36, 155
  composition of, 36
    regular glaze, 37
    ion exchange glazes, 159

Glaze [Cont.]
  thermal expansion of, 36,
    157
  thickness of, 45, 64, 155
Glazing, 14, 36-80, 155-162
  of alumina, 155-162
  of forsterite, 126
  of glass ceramics, 155
  ion exchange, 14, 158-162
  low expansion, 155
  of sapphire, 111, 113
  of spinel, 121
  of steatite, 125
  of titania, 121
  of zircon porcelain, 128
Glazing and quenching, 14,
    36-80
  of alumina, 36
  delayed fracture, 65-72
  of electrical porcelain, 128
  flexural strength, 43,
    52, 55
    abraded, 44
    elevated temperature, 58
  of forsterite, 126
  of sapphire, 111, 114, 116
  of spinel, 12
  of steatite, 124
  thermal shock resistance, 64
  of titania, 121
  of zircon porcelain, 128
Grain size, 8
Griffith's theory, 3

                  H

Hackle, 15, 92, 97, 109
Heating  (see Refiring)

                  I

Impact resistance, 13, 28, 31
  of alumina, 60-63
    elevated temperature, 62
  Charpy test, 28
  drop weight test, 28

Impact resistance [Cont.]
  of glass-ceramic, glazed
    cylinders, 155
  of silicon carbide, 132, 135
Indentation test, 23, 74
Ion exchange glazing, 14
  of alumina, 158
  composition profile, 163

                  L

Leaching, 170
  of alumina, 170
  of spinel, 179
Limitations of treatments,
    215
Load, 18
Localized stress, 7, 9

                  M

Machining, 73
Magnesia, 182
  MgO-NiO solid solution on,
    182-186
  forsterite on, 203
Mullite (see Aluminosilicate
    ceramic)
  quenching, 128, 131
  on sapphire, 199
  on sintered alumina, 195

                  N

Nickel oxide, 187
Notch, 3

                  P

Penetration, of damage,
    29, 31, 91-95
Parasitic stress, 26
Periclase (MgO crystal), 186

Periclase [Cont.]
  MgO-NiO solid solution on,
    186
  MgO-MnO solid solution on,
    186
Phase transformation, 14
  of zirconia, 205-207
Property measurements, 20


                    Q

Quenching, 14, 18, 19
  of alumina, hot pressed,
    81-110
  of alumina, sintered, 31-80
  of aluminosilicate ceramic,
    51, 54
  benefits of, 153
  delayed fracture, 66-72
  variables
    grain size, 86
    indentation, 74
    quenching conditions, 83
    scratching, 74, 91-96
    starting materials, 83
    temperature, 51, 54
  of electrical porcelain, 128,
    130
  flexural strength, 32, 82
    elevated temperature, 58, 86
  impact resistance, 60-63
  media, 31, 45, 49
    air blast (forced air), 31,
      36, 39, 53, 121, 124
    emulsions, 32, 34
    gaseous media, 31
    fluidized bed, 135
    liquid media, 31, 82, 124
  of other oxides and silicates,
    120-131
  of mullite, 128, 131
  of sapphire, 110-120
  of silicon carbide, 132
  of silicon nitride, 139-143
  of spinel, 121
  of steatite, 124
  thermal shock resistance, 63-65
  of titania, 121
  of zircon porcelain, 128

                    R

Reaction, 14, 188
Recommendations for research,
    154, 212
  thermal treatments, 154
  coatings, chemical treatments
    and phase transformations,
    212
Refiring, 54
  of alumina, 54, 58, 62
  of sapphire, 110
  of spinel, 121
  of silicon carbide, 132
  of silicon nitride, 137
Reflecting spots, 104, 106,
    108
Reheating (see Refiring)
Ring test, 20, 38, 170, 172,
    177, 180
Rod test, 21, 38, 52
  on alumina
    glazed, 40, 156
    hot pressed, 83
    ion exchange glazed,
      163
    96%, 52, 191, 195
    sintered, 199
  on sapphire, 111, 173,
    190, 199
  on silicon carbide, 134
  on zirconia, 205


                    S

Sapphire, 110
  chemical polishing, 110,
    115
  chemical treatments, 173-176,
    199-203
  flame polishing, 110, 115
  flexural strength of, 119-116
  glazing of, 113
  glazing and quenching of,
    111-114
  quenching of, 110
  thermal shock resistance of,
    117-120
  thermal treatments of, 110-120

Scratching, 73
Selection of bodies for
    strengthening, 14
Shape and size, 148
  effect of, in thermal
    treatments, 148-153
Silica, 1, 206
Silicon carbide, 132
  quenching of, 134-137
  refiring of, 132-137
Silicon nitride, 137-145
  quenching of, 139-145
  refiring of, 137-139
  SIALON solid solutions on,
    188
Slotted rod test (see Rod test)
Solid solution surface layers,
    14, 163-188
  Al$_2$O$_3$-Cr$_2$O$_3$, 163
    halide additions, 171
    on alumina, 167-173
    on sapphire, 173-176
    thermal expansion of,
      163, 165
  MgAl$_2$O$_4$-MgCr$_2$O$_4$, 177-182
    on spinel, 177-182
  MgO-NiO, 186-187
    on magnesia, 186
    on nickel oxide, 187
  SIALON, 188
    on silicon nitride, 188
  stress analysis, 166
  TiO$_2$-SnO$_2$, 176
    on titania, 176
Spinel, 120, 123
  MgAl$_2$O$_4$-MgCr$_2$O$_4$, solid
    solution, 177-182
  glazing of, 123
  glazing and quenching of,
    123
  other surface treatments,
    196
Static fatigue (see Delayed
    fracture)
Status of research, 145-154,
    208-214
  coatings, chemical treatments
    and phase transformations,
    208-214
  thermal treatments, 145-154

Steatite, 120, 124
  glazing of, 124
  glazing and quenching of,
    124
  quenching of, 125
Strain, 19
  holographic interferometry,
    71
Strength, 1
  comparison of tensile and
    flexural, 27
  degradation, 73, 91-96, 117
  distributions, 11, 27, 54,
    58, 60, 67, 217
    alumina, refired, 54-58
  at elevated temperature,
    56, 60, 86
  flexural, 13, 24, 27, 32, 35
    of alumina, hot pressed,
      82, 91, 100, 194
    of alumina, 96%, 40, 41,
      43, 46, 51-60, 152, 156,
      160, 169, 193, 198
    of alumina, sintered,
      191-199
    of aluminosilicate ceramic,
      54
    of electrical porcelain, 128
    errors, 25
    of forsterite, 126
    of magnesia, 184
    measurement of, 24
    of nickel oxide, 187
    of sapphire, 113, 116, 175,
      201
    of silicon carbide, 137
    of silicon carbide, 137
    of silicon nitride, 137, 140
    of spinel, 180
    of steatite, 125
    of titania, 121, 177
    of zircon porcelain, 128
    of zirconia, 206
  tensile, 26, 45
    of alumina, 49, 162
    of silicon nitride, 141, 143
  units of, 26
Strengthening, 12, 14
  mechanisms, 12
  technique, 14

Stress, 1
  axial, 17, 82, 90, 101
  circumferential, 17, 90
  compressive surface, 13, 165
  concentrations, 1, 2
  corrosion mechanisms, 72
  distribution, 3
  fracture, 24
    units, 26
  intensity factor, 5, 9
    crack branching, 23
  internal, 18
  localized, 7
  parasitic, 26
  profile, 17- 56
    of alumina, hot pressed,
      95-103
    of alumina, 96%, 56, 74-80,
      167
    of magnesia, 186
  radial, 17
  relaxation, 19, 55, 57,
    113
  residual, 18, 22, 24, 45
    of alumina, hot pressed,
      98-102
    fracture mirror measurement
      of, 98-102, 141
    indentation measurement
      of, 23
    weakening as a result of,
      89
    x-ray measurement of, 78
  reversal, 19
Subcritical crack growth,
    5, 9, 216
  boundary, 11
Successes and failures, 148
  coatings, chemical treatments
    and phase transformations,
    209
  thermal treatments, 148
Surface, 4
  damage, 13, 29, 31, 218
    abrasion, 42, 73, 115
    scratching, 73, 91-95
  energy, 4
  relative force measurements,
    20
  layer thickness, 17, 45

T

Thermal
  conditioning, 31
  expansion coefficients,
    13, 17
    of alumina, 36
    of $Al_2O_3$-$Cr_2O_3$, solid
      solution, 165
    of glazes, 36, 155, 157
    of $MgAl_2O_4$-$MgCr_2O_4$, solid
      solution, 178
    of MgO-NiO solid
      solution, 182
    mismatch, 36
  of NiO, 187
  of $TiO_2$-$SnO_2$ solid solution,
    165
  shock resistance, 13, 29, 31
    of alumina, 63, 171
    of aluminosilicate ceramic,
      51, 54
    of sapphire, 117, 174, 201
    of spinel, 181
  treatments, 31-154
Titania, 120
  fluorine treatments of, 121
  glazing and quenching of, 121
  quenching of, 121
  $TiO_2$-$SnO_2$ solid solutions on,
    176

UVW

Units, 26
Volume changes, 189
Weibull distribution, 17
Weight reduction, 220

Z

Zircon porcelain, 120
  glazing of, 128
  glazing and quenching of,
    128
  quenching of, 128
Zirconia, 205
  phase transformation,
    205